A Book Of

ATOMIC AND MOLECULAR PHYSICS

T.Y.B.Sc. Physics : PH - 334 : Semester-III

As Per New Revised Syllabus with Effect from June 2015

Dr. P. S. TAMBADE
M.Sc., Ph.D.
Department of Physics,
Prof. Ramkrishna More,
Arts, Commerce & Science College,
Akurdi Pradhikaran, **PUNE**.

Dr. S. D. AGHAV
M.Sc., M.Phil, Ph.D.
Ex. Vice Principal
Bahuraoji Gholap Mahavidyalaya,
Sangvi, **PUNE**

Dr. G. R. PANSARE
M.Sc., Ph.D.
H. V. Desai College,
PUNE.

B. M. LAWARE
M.Sc., M. Phil.
Head, Department of Physics,
Prof. Ramkrishna More,
Arts, Commerce & Science College,
Akurdi Pradhikaran, **PUNE**.

V. K. DHAS
M.Sc., M. Phil.
Ex. Head, Department of Physics,
New Arts, Commerce & Science College,
AHMEDNAGAR.

Dr. B. G. WAGH
M.Sc., Ph.D.
Principal
KSKW Arts, Science and Commerce College,
CIDCO, **NASHIK**.

T.Y.B.Sc. ATOMIC AND MOLECULAR PHYSICS   **ISBN 978-93-51645-88-7**

Fifth Edition   :   August 2019

©   :   **Authors**

Published By :
NIRALI PRAKASHAN
Abhyudaya Pragati, 1312, Shivaji Nagar,
Off J.M. Road, PUNE – 411005
Tel - (020) 25512336/37/39, Fax - (020) 25511379
Email : niralipune@pragationline.com

➤ DISTRIBUTION CENTRES

PUNE

Nirali Prakashan : 119, Budhwar Peth, Jogeshwari Mandir Lane, Pune 411002, Maharashtra
(For orders within Pune)   Tel : (020) 2445 2044, Mobile : 9657703145
Email : niralilocal@pragationline.com

Nirali Prakashan : S. No. 28/27, Dhayari, Near Asian College Pune 411041
(For orders outside Pune)   Tel : (020) 24690204 Fax : (020) 24690316; Mobile : 9657703143
Email : bookorder@pragationline.com

MUMBAI

Nirali Prakashan : 385, S.V.P. Road, Rasdhara Co-op. Hsg. Society Ltd.,
Girgaum, Mumbai 400004, Maharashtra; Mobile : 9320129587
Tel : (022) 2385 6339 / 2386 9976, Fax : (022) 2386 9976
Email : niralimumbai@pragationline.com

➤ DISTRIBUTION BRANCHES

JALGAON

Nirali Prakashan : 34, V. V. Golani Market, Navi Peth, Jalgaon 425001, Maharashtra,
Tel : (0257) 222 0395, Mob : 94234 91860;
Email : niralijalgaon@pragationline.com

KOLHAPUR

Nirali Prakashan : New Mahadvar Road, Kedar Plaza, 1st Floor Opp. IDBI Bank, Kolhapur 416 012
Maharashtra. Mob : 9850046155; Email : niralikolhapur@pragationline.com

NAGPUR

Nirali Prakashan : Above Maratha Mandir, Shop No. 3, First Floor,
Rani Jhanshi Square, Sitabuldi, Nagpur 440012, Maharashtra
Tel : (0712) 254 7129; Email : niralinagpur@pragationline.com

DELHI

Nirali Prakashan : 4593/15, Basement, Agarwal Lane, Ansari Road, Daryaganj
Near Times of India Building, New Delhi 110002 Mob : 08505972553
Email : niralidelhi@pragationline.com

BENGALURU

Nirali Prakashan : Maitri Ground Floor, Jaya Apartments, No. 99, 6th Cross, 6th Main,
Malleswaram, Bengaluru 560003, Karnataka; Mob : 9449043034
Email: niralibangalore@pragationline.com

Other Branches : Hyderabad, Chennai

niralipune@pragationline.com | www.pragationline.com
Also find us on   www.facebook.com/niralibooks

Preface ...

The present book entitled **"Atomic and Molecular Physics"** is written as per new revised syllabus prescribed for the IIIrd Semester of T.Y.B.Sc. (Physics) of Savitribai Phule Pune University from June 2015. This book is targeted mainly to the undergraduate students of Pune University, but will be found useful for the graduate students and Teachers of other universities also. The book is divided into 7 chapters. Each chapter begins with basic concepts containing theory, set of formulae and explanatory notes for followed by a number of solved problems. Summary of contents in topic, short, long questions and unsolved problems are also given at the end of each topic.

The problems are judiciously selected and are given topic and section-wise. The approach is straight forward and step-by step solutions are elaborately provided. More importantly the relevant formulas used for solving the problems can be located in the beginning of each chapter. There are number of diagrams for illustration.

Chapter 1 in the book is devoted to Atomic Structure. Chapter 2 is basically concerned One Valence Electron Systems. Chapter 3 is concerned with Two Valence Electron Systems. Chapter 4 is basically related to Zeeman Effect. Chapter 5 is related to X-Ray Spectroscopy. Chapter 6 is concerned with Molecular Spectroscopy and Chapter 7 dealt with Raman Spectroscopy.

All precautions have been taken to avoid mistakes and misprint in the book. However, it is possible that some mistakes and misprints might have passed unnoticed. Such mistakes and misprint, is brought to our notice will be thankfully acknowledged.

We are thankful to Shri Jignesh Furia and staff of Nirali publication for publishing the book in attractive look. We have a pleasure to thank Mr. Santosh Bare for the bulk of typing and Mr. Kiran Velankar for proof reading. I am indebted to Mrs. Anjali Muley for line drawings, to Ravi Walodare for designing cover page and all staff in the distribution of books network.

We are also thankful to all the Marketing Staff especially Mr. Nilesh Deshmukh and others for co-ordinating the matter well in time.

Suggestions to improve the quality of the book will be gladly accepted.

AUTHORS

Syllabus ...

1. Atomic Structure (6 L)

1. Rutherford's model of atom
2. Electron orbits
3. Bohr atom
4. Energy levels and spectra (1 to 4 Revision)
   Vector atom model (Concepts of space and quantization and electron spin)
5. Atomic excitation and atomic spectra, Problems.

2. One and Two Valence Electron Systems (7 L)

1. Pauli Exclusion principle and electron configuration,
   Quantum states, Spectral notations of quantum states.
2. Spin-Orbit Interaction (Single valence electron atom), Energy levels of
   Na atom, Selection rules, Spectra of sodium atom, Sodium doublet.

3. Two Valence Electron Systems (7 L)

1. Spectral terms of two electron atoms, Terms for equivalent electrons,
   LS and jj coupling schemes.
2. Singlet-Triplet separation for interaction energy of LS coupling.
   Lande's Interval rule, Spectra of Helium atom, Problems.

4. Zeeman Effect (4 L)

1. Early discoveries and developments
2. Experimental arrangement
3. Normal and anomalous Zeeman Effect problems
4. Stark effect (Qualitative discussion)

5. X-Ray Spectroscopy (6 L)

1. Nature of X-rays
2. Discrete and continuous X-ray spectra, Duane and Hunt's rule
3. X-ray emission spectra
4. Moseley's law and its applications
5. Auger effect, Problems.

6. Molecular Spectroscopy (10 L)

1. Rotational energy levels
2. Vibrational energy levels
3. Rotational and Vibrational spectra
4. Electronic spectra of molecules, Problems.

7. Raman Spectroscopy (8 L)

1. Classical theory of Raman Effect - Molecular polarizability
2. Quantum theory of Raman Effect
3. Experimental Set up for Raman Effect
4. Applications of Raman Spectroscopy.

•••

Contents ...

❑❑❑

Chapter **1**...

Atomic Structure

Contents ...

James Franck

James Franck (26 August 1882 – 21 May 1964) was a German physicist and Nobel laureate (with Hertz).

In 1914, James Franck and Gustav Hertz performed an experiment which demonstrated the existence of excited states in mercury atoms, helping to confirm the quantum theory which predicted that electrons occupied only discrete, quantized energy states.

Introduction

• Every atom consists of nucleus of proton and neutron with number of electrons revolving around the nucleus. It was thought that electrons revolve around the nucleus like planets do around the sun. But classical electromagnetic theory rejected possibility of stable electron orbit. To resolve this paradox Neil Bohr applied quantum theory to understand atomic structure in 1913 and developed a model which is still a convenient mental picture of an atom.

• In this chapter, the chief concern will be on Rutherford model, electron orbits, Bohr atom and energy levels. In addition vector atom model and concept of atomic excitation is discussed.

1.1 Rutherford's Model of an Atom

- Discovery of electron by J.J. Thomson provided a starting point for theories of atomic structure. J.J. Thomson assumed that an atom consisted of electron distributed in the positively charged sphere of radius 10^{-10} m. This model could not explain all features of optical spectra of hydrogen and other elements and hence the theory was discarded.

- At the suggestion of Ernest Rutherford two scientists Hans Geiger and Ernest Marsden used as probes for the fast alpha particles emitted by certain radioactive elements. Alpha particles are helium atoms that have lost two electrons each leaving them with a charge of + 2e.

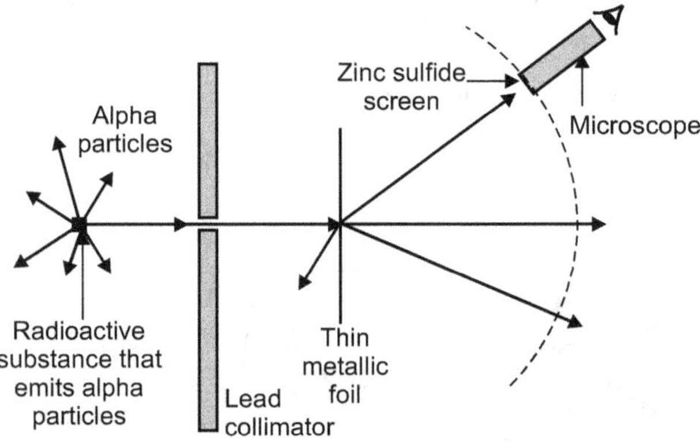

Fig. 1.1 : Rutherford's scattering experiment

- They placed a sample of a alpha-emitting substance behind a lead screen with a fine hole as shown in Fig. 1.1 so that narrow beam of alpha particles was produced. The beam was directed at a thin gold foil. Screen was made up of zinc sulphide. Screen gives off a visible flash of light when struck by alpha particle was set on the other side of the foil with a microscope to observe flashes.

- It was expected that alpha particle would go right through foil without any deviation. Geiger and Marsden observed that most of the alpha particles were not deviated by much but a few were scattered through very large angles. Some of them were even scattered in backward direction. Alpha particles are relatively heavy ($\approx$ 8000 times mass of electron) and have a speed of 2×10^7 m/s, so it was clear that powerful forces are there which cause such remarkable deflection.

- To explain this result Rutherford put forward his model. According to the model, an atom is composed of tiny nucleus in which its positive charge and nearly all its mass are concentrated with electron some distance away. (Refer Fig. 1.2)

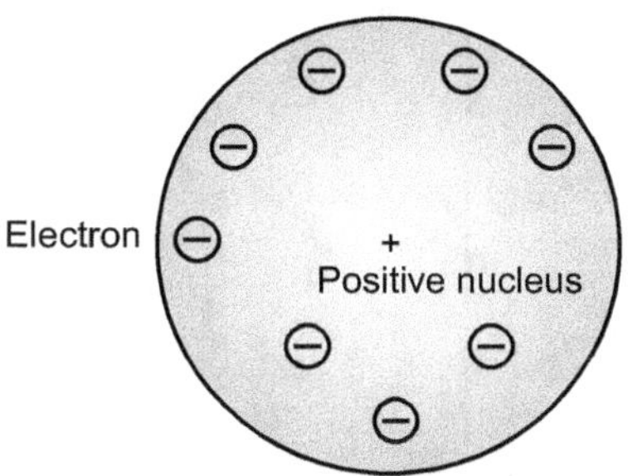

Fig. 1.2 : Rutherford's model of an atom

- When an alpha particle happens to come near a nucleus, the intense electric field there scatters it through large angle. Being electrons are light, they do not appreciably affect the alpha particles. It is also observed that the deflection of an alpha particle when it passes near nucleus depends on the magnitude of the nuclear charge. Comparing the relative scattering of alpha particles by different foils provides information about charges of an atom involved. The nuclear charges are to be multiple of +e and the number Z (called atomic number of element). We know that proton, each with charge +e, provides the charge on a nucleus, so that the atomic number of an element is the same as number of protons in the nuclei of its atom.

- The formula that Rutherford obtained for alpha particle scattering by a thin foil on the basis of nuclear model is

$$N(\theta) = \frac{N_i \, ntZ^2 \, e^4}{(8\pi\varepsilon_o)^2 \, r^2 \, KE^2 \, \sin^4(\theta/2)} \qquad \ldots (1.1)$$

where　　$N(\theta)$ = Number of alpha particles per unit area that reach the screen at scattering angle θ

N_i = Total number of alpha particles that reach the screen

n = Number of atoms per unit volume in the foil

Z = Atomic number of the foil atom

r = Distance of the screen from the foil

KE = Kinetic energy of the alpha particle

t = Foil thickness

- Prediction of equation (1.1) agrees with measurements of Geiger and Marsden. As $N(\theta) \propto \dfrac{1}{\sin^4(\theta/2)}$, the variation of $N(\theta)$ with θ is pronounced. Refer Fig. 1.3. Only 0.14% of incident alpha particles are scattered by more than 1°.

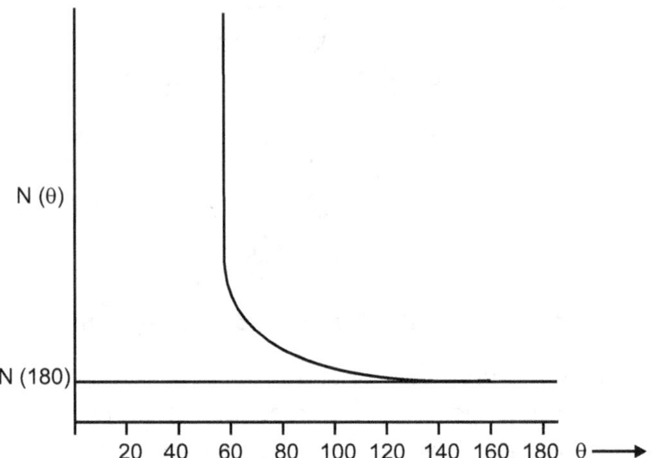

Fig. 1.3 : Rutherford's scattering N(θ) versus θ

- In the derivation of equation (1.1), Rutherford assumed that size of target nucleus is small compared with the minimum distance R to which incident alpha particles approach the nucleus before being deflected away. Thus, Rutherford's scattering gives us a way to find an upper limit of nuclear dimension.

- Let us calculate distance of closest approach R for an alpha particle. Alpha particle will have smallest R when it approaches to a nucleus head on, which will be followed by a 180° scattering. At the instant of closest approach, the initial kinetic energy KE of particle is entirely converted to electric P.E. so that

$$\text{K.E. (initial)} = \text{P.E.} = \frac{1}{4\pi\varepsilon_o}\frac{2\,Ze^2}{R}$$

Charge on α particle = 2e and that of nucleus = Ze

$\therefore$ Distance of closest approach,

$$R = \frac{2\,Ze^2}{4\pi\varepsilon_o\,KE_{initial}} \qquad \text{... (1.2)}$$

In natural origin, max. K.E. = 7.7 MeV = 1.2×10^{-12} J

Taking $\dfrac{1}{4\pi\varepsilon_o} = 9 \times 10^9$ Nm²/l^2

$$R = \frac{9 \times 10^9 \times (1.6 \times 10^{-19})^2}{1.2 \times 10^{-12}}\,Z$$

$$= 3.8 \times 10^{-16}\,Z \ \text{meter}$$

For gold, Z = 79.   R (AU) = 3×10^{-14} m

- Radius of gold nucleus is less than 3×10^{-14}, well under 10^{-4} the radius of an atom as a whole.

1.2 Electron Orbits

- Rutherford's model confirms model of an atom. According to him an atom is tiny, massive, positively charged nucleus surrounded by electrons at relatively great distance to make an atom electrically neutral as whole.

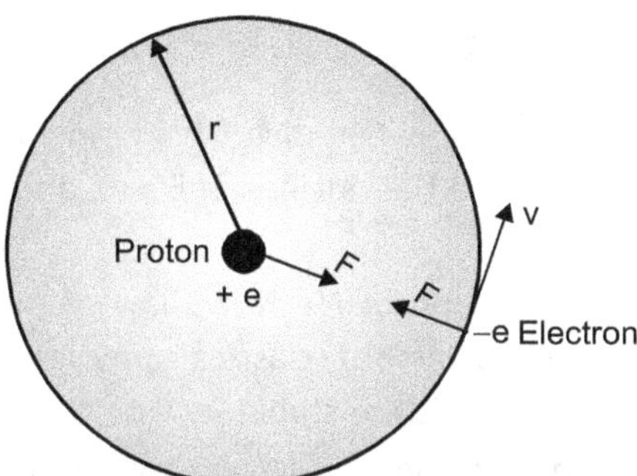

Fig. 1.4 : Force balance in a hydrogen atom

- Consider the case of a simple hydrogen atom having a single electron. We assume a circular electron orbit for convenience. The centripetal force holding electron in an orbit of radius r is given by

$$F_c = \frac{mv^2}{r}$$

Similarly, nucleus is provided by electric force.

$$F_e = \frac{1}{4\pi\varepsilon_o}\frac{e^2}{r^2}$$

Condition for a dynamic stable orbit is

$$F_c = F_e$$

$$\frac{mv^2}{r} = \frac{1}{4\pi\varepsilon_o}\frac{e^2}{r^2}$$

Hence velocity v related to orbital radius is

$$v = \frac{e}{\sqrt{4\pi\varepsilon_o\, mr}} \qquad \text{... (1.3)}$$

- Total energy of hydrogen atom is the sum of kinetic and potential energies.

$$K.E. = \frac{1}{2}mv^2 \quad \text{and} \quad P.E. = -\frac{e^2}{4\pi\varepsilon_o\, r}$$

$$E = K.E. + P.E. = \frac{mv^2}{r} - \frac{e^2}{4\pi\varepsilon_o\, r}$$

Substituting the value of v from equation (1.3),

$$E = \frac{e^2}{8\pi\varepsilon_o\, r} - \frac{e^2}{4\pi\varepsilon_o\, r}$$

Total energy of an hydrogen atom

$$E = -\frac{e^2}{8\pi\varepsilon_o\, r} \qquad\qquad \ldots (1.4)$$

- Total energy of an hydrogen atom is negative. This holds for every atomic electron and reflects the fact that it is bound to the nucleus. If E > 0, and electron would not follow a closed orbit around nucleus.

- The above analysis is an application of Newton's law of motion and Coulomb's law of electric force. Both laws are pillars of classical physics and is in accord with the experimental observation that atoms are stable.

- But it is not in accord with electromagnetic theory, another pillar of classical theory. According to electromagnetic theory the accelerating electric charges radiate energy in the form of wave. An electron pursuing a curved path is accelerated and therefore should continuously lose energy into nucleus in a fraction of second. (Refer Fig. 1.5) But atoms do not collapse. The laws of physics that are valid in macroworld do not hold true in the microworld of atom.

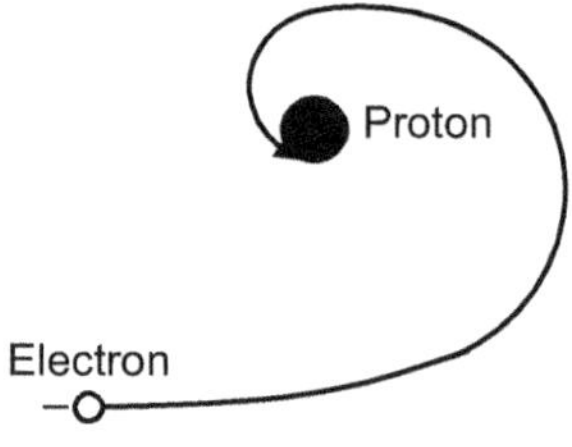

Fig. 1.5

- Classically atomic electron spiral rapidly into the nucleus as it radiates energy due to its acceleration.

- Thus, classical physics fails to provide a meaningful analysis of atomic structure because it approaches the nature in terms of pure particles and pure waves. In reality, particles and waves have many properties in common though the smallness of Planck's constant make the wave particle duality imperceptible in the macroworld. The usefulness of classical physics is less for microworld, we must allow particle behaviour of waves and wave behaviour of particle to understand atom. In remaining sections of the topic, we will see how the Bohr's atomic model which combines classical and modern approach.

1.3 Bohr Atom

- The hydrogen atom theory proposed by Neil Bohr in 1913, marked the beginning of a new era in atomic structure and spectroscopy. Bohr's theory provided the satisfactory explanation of the Balmer, Lyman, and Paschen series of hydrogen atom and the pickering series of ionized helium. In addition, he calculated the Rydberg constant on the basis of purely theoretical considerations.

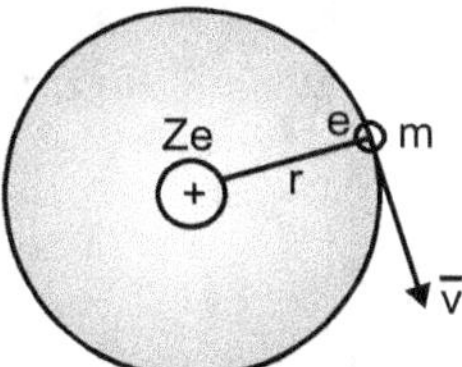

Fig. 1.6 : Bohr's circular orbit

- In short, this theory gave a physical meaning to the experimental discovery that the frequency of spectral line is given by the difference between two terms.

- Bohr modified the Rutherford's model by using the Planck's quantum theory of radiation. He assumed that the hydrogen atom consists of a stationary nucleus (proton) at the centre of atom whose mass is 1836 times mass of electron. A single electron revolves around the stationary nucleus. According to the Planck's quantum theory (1900), energy is radiated and absorbed in whole multiples of an elementary quantum of energy,

$$E = h\upsilon \qquad \qquad \text{... (1.5)}$$

where, υ is the frequency of the emitted or absorbed radiation and h is Planck's constant of action (h = 6.625×10^{-34} J.s). This quantum of energy is called photon.

Bohr's theory is based on the following three postulates.

First Postulate :   **(April 12)**

- Bohr's first assumption was that an *electron revolves in certain allowed circular orbits about the nucleus under the action of Coulomb's force.* The force of attraction between the electron of charge e and the nucleus of charge Ze is given by

$$F = \frac{1}{4\pi\varepsilon_0}\frac{Ze^2}{r^2} \qquad \qquad \text{... (1.6)}$$

where a is the electron-nuclear distance, Z is the atomic number (Z = 1 for hydrogen) and ε_0 is the permittivity of free space. The force given by equation (1.6) is equal to the centripetal force mv^2/a, where v is the velocity and m is the mass of the electron. For equilibrium condition,

$$\frac{mv^2}{r} = \frac{1}{4\pi\varepsilon_0}\frac{Ze^2}{r^2} \qquad \qquad \text{... (1.7)}$$

which gives

$$\boxed{v^2 = \frac{1}{4\pi\varepsilon_0}\cdot\frac{Ze^2}{mr}} \qquad \qquad \text{... (1.8)}$$

Second Postulate :　　　　　　　　　　　　　　　　　　　　**(Oct. 16, April 16)**

* Bohr's second assumption for the hydrogen atom may be stated as "*only those circular orbits for electrons are permitted for which the angular momentum $(\bar{L}\,)$ is equal to the integral multiple of $\dfrac{h}{2\pi}$.*

Thus,
$$\bar{L} = n\left(\frac{h}{2\pi}\right) \qquad \dots (1.9)$$

where, n = 1, 2, 3, ...

* Orbital angular momentum of an electron is given by

$$\bar{L} = mvr \qquad \dots (1.10)$$

where v is the linear velocity of electron in an orbit.

Equations (1.9) and (1.10) give

$$\boxed{mvr = n\left(\frac{h}{2\pi}\right)} \qquad \dots (1.11)$$

* Orbits permitted by the condition shown by equation (1.7) are called allowed orbits for electrons. These orbits are called non-radiating orbits.

Third Postulate :　　　　　　　　　　　　　　　　　　　**(April 17; Oct. 17, 11)**

* *A quantum of energy is radiated only when an electron jumps from higher energy orbit to the lower energy orbit and the frequency of emitted radiation is proportional to the difference between two energy states.*

Thus,
$$h\upsilon = E_1 - E_2 \qquad \dots (1.12)$$

where υ is the frequency, E_1 and E_2 are the energies of initial and final states.

From equation (1.7), we get

$$mv^2 = \frac{1}{4\pi\varepsilon_0}\frac{Ze^2}{r}$$

$$m^2v^2 = \frac{1}{4\pi\varepsilon_0}\frac{Ze^2m}{r} \qquad \dots (1.13)$$

Equation (1.11) gives

$$mv = \frac{n}{r}\frac{h}{2\pi}$$

$$m^2v^2 = \frac{n^2}{r^2}\frac{h^2}{4\pi^2} \qquad \dots (1.14)$$

From equations (1.13) and (1.14), we get

$$r = \frac{\varepsilon_0\, n^2\, h^2}{\pi m e^2 Z} \qquad \dots (1.15)$$

For hydrogen atom, $Z = 1$. So the radius of nth orbit is given by

$$\boxed{r_n = \frac{\varepsilon_0 \, n^2 \, h^2}{\pi m e^2}} \qquad \ldots (1.16)$$

where n is called principal quantum number.

Equation (1.16) indicates that

$$r_n \propto n^2, \quad n = 1, 2, 3, \ldots\ldots$$

Thus radii of Bohr's orbit are proportional to the square of natural numbers.

Radius of first orbit, $n = 1$ is

$$r_1 = 0.53 \times 10^{-10} \, m = 0.53 \, A°$$

From equations (1.11) and (1.16), the velocity of an electron in n^{th} orbit is given by

$$\boxed{v_n = \frac{e^2}{2\varepsilon_0 n h}} \qquad \ldots (1.17)$$

1.4 Energy Levels and Spectra　　　　　　　　　　　(Oct. 17)

- The kinetic energy of an electron in n^{th} orbit will be

$$T = \frac{1}{2} m \, v_n^2 \qquad \ldots (1.18)$$

$$\therefore \qquad \boxed{T = \frac{m e^4}{8 \varepsilon_0^2 \, n^2 \, h^2}} \qquad \ldots (1.19)$$

- The potential energy (U) of the electron, when it is at a distance 'a' from the nucleus, is given as

$$U = \int_a^\infty F \cdot dr \qquad \ldots (1.20)$$

where F is the Coulomb's electrostatic force given by equation (1.6). Thus equation (1.20) gives

$$U = -\frac{m e^2}{4 \varepsilon_0^2 \, n^2 \, h^2} \qquad \ldots (1.21)$$

The total energy of an electron in n^{th} orbit is sum of kinetic and potential energies.

Thus equations (1.19) and (1.21) give,

$$E_n = T + U$$

$$\boxed{E_n = -\frac{m e^4}{8 \varepsilon_0^2 \, n^2 \, h^2}} \qquad \ldots (1.22)$$

where E_n is the total energy of an electron in n^{th} orbit.

Energies specified by above equation are called energy levels.

Equation (1.22) shows that

$$\boxed{E_n \propto \left(\frac{1}{n^2}\right)}$$

For the first orbit, n = 1, and substituting values in above equation, we get

$$E_1 = -2.18 \times 10^{-18} \text{ J}$$

$$E_1 = -13.6 \text{ eV}$$

- Thus 13.6 eV energy is required to remove the electron completely from the first orbit of hydrogen atom. This is called the **binding energy** of an electron in first orbit. This is the minimum energy state of an electron. This state is called the **ground state.** The energy of an electron in the subsequent states is given by,

$$\boxed{E_n = -\frac{13.6}{n^2} \text{ eV}} \qquad \text{... (1.23)}$$

- Equation (1.23) indicates that the total energy of an electron increases as it goes away from the nucleus. This also indicates that energy difference of electron in consecutive orbits decreases with increase in the value of n.

Origin of line spectra :

- According to the third postulate, when there is transition of an electron from higher energy orbit to the lower energy orbit, a photon of energy $h\upsilon$ is emitted. Let us consider the transition of electron from higher energy n^{th} orbit to lower energy m^{th} orbit, then the energy quanta emitted is given by,

$$h\upsilon = E_n - E_m \qquad \text{... (1.24)}$$

where E_n is the energy of an electron in n^{th} orbit and E_m is the energy of an electron in m^{th} orbit.

- From equations (1.22) and (1.24), we get

$$h\upsilon = -\frac{me^4}{8\varepsilon_0^2 h^2 n^2} - \left[-\frac{me^4}{8\varepsilon_0^2 h^2 m^2}\right]$$

$$\upsilon = \frac{me^4}{8\varepsilon_0^2 h^3}\left[\frac{1}{m^2} - \frac{1}{n^2}\right] \qquad \text{... (1.25)}$$

But

$$c = \upsilon\lambda$$

where λ is wavelength of spectral line, c is the velocity of light.

Thus from equation (1.25), we get

$$\frac{1}{\lambda} = \frac{me^4}{8\varepsilon_0^2 ch^3}\left[\frac{1}{m^2} - \frac{1}{n^2}\right] \qquad \text{... (1.26)}$$

The quantity $\frac{1}{\lambda} = \bar{v}$ is called wave number.

∴
$$\bar{v} = R\left[\frac{1}{m^2} - \frac{1}{n^2}\right]$$
... (1.27)

where R is the Rydberg's constant and it is given by

$$R = \frac{me^4}{8\varepsilon_0^2\, ch^3} = 1.097 \times 10^7 \text{ m}^{-1}$$
... (1.28)

Hydrogen spectrum consists of the following five series of spectral lines.

(1) Lyman series : This series exists when there is transition of an electron from initial orbit n = 2, 3, 4 ... to final lower orbit m = 1. Wavelength of spectral lines of this series lies in the ultra-violet region.

(2) Balmer series : Spectral lines of Balmer series are observed when an electron transition is from initial orbit n = 3, 4, 5 ... to final lower orbit m = 2. This gives the visible region.

H_α line : transition of electron is from n = 3 to m = 2

H_β line : transition of electron is from n = 4 to m = 2.

H_γ line : transition of electron is from n = 5 to m = 2.

H_δ line : transition of electron is from n = 6 to m = 2

and so on ....

(3) Paschen series : For this series, electron transition is from n = 4, 5, 6 ... to m = 3. Spectral lines lie in the near infrared region.

(4) Bracket series : For this series, electron transition is from n = 5, 6, 7 ... to m = 4. Spectral lines lie in the infrared region.

(5) Pfund series : For this series, electron transition is from n = 6, 7, 8 ... to m = 5. It lies in the far infrared region.

The spectral wavelengths calculated using the Bohr's relation (1.27) are in good agreement with the experimentally observed wavelengths for hydrogen atom. This is the good success of Bohr's theory.

Limitations of Bohr's Theory :	**(Oct. 17, April 16)**

Following are few limitations of Bohr's theory.

1.	It cannot explain the relative intensities of the spectral lines.

2.	It cannot explain the spectra of two or more electron atom.

3.	It cannot give details of fine structure of spectral line. According to the Bohr's theory, H_α line is single line of wavelength 6563 A°. But experimentally it is observed that H_α line consists of two spectral lines of small wavelength difference 1.4 A°. Bohr's theory could not explain this fine structure of spectral line.

Energy Levels and Series Transitions of Hydrogen :   **(Dec. 11)**

- Fig. 1.7 shows the energy levels and different series transitions of hydrogen atom. In an energy level diagram, each Bohr orbit or stationary state is represented by a horizontal line. Vertical scale is taken in terms of wave number $(\bar{v})$. Equation (1.22) shows that the energy values of quantized states are negative and approach to zero as the principal quantum number n increases.

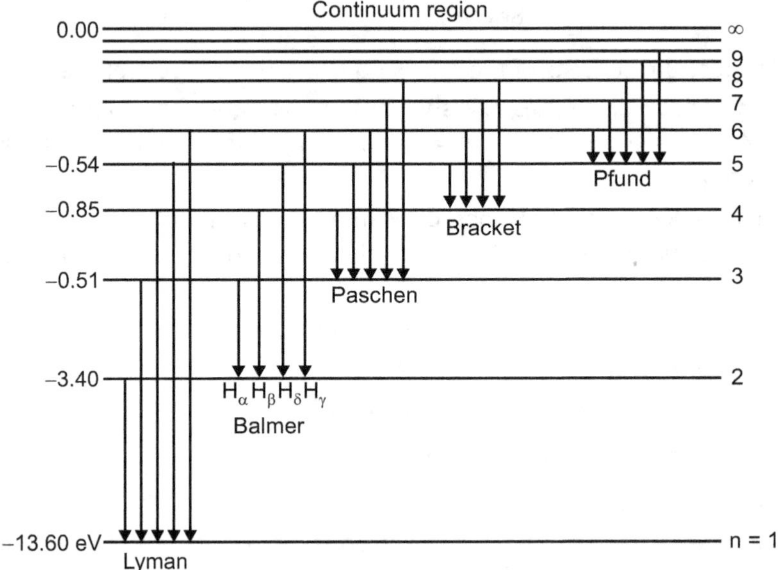

Fig. 1.7 : Energy levels and series transitions of hydrogen atom (Not to the scale)

Bohr-Sommerfeld's Theory for Hydrogen like Atom :

- Bohr's theory of hydrogen atom gives the satisfactory explanation for the existence of different spectral lines. Results of Bohr's theory are in good agreement with the experimental observations for the hydrogen atom. But this theory fails to explain the fine structure of atom and the different relative intensities of spectral lines. This may be due to Bohr's assumptions. He assumed that the nucleus is stationary and an electron moves in the circular orbit around the nucleus. Mass of an electron is considered to be negligible in comparison with the nucleus.

- Sommerfeld considered the motion of an electron and the nucleus such that both move around their common centre of mass. This gives the concept of reduced mass of an electron-nucleus system as a whole.

- According to Sommerfeld, electron revolves in an elliptical orbit with the nucleus at one of the focus. The position of electron at any time instant can be specified by polar co-ordinates r and θ (Refer Fig. 1.8). In case of circular orbit, only one co-ordinate θ varies (length of radius vector constant). However, in the case of elliptical orbit not only angle θ varies but the length of the radius vector r varies periodically as shown in Fig. 1.8.

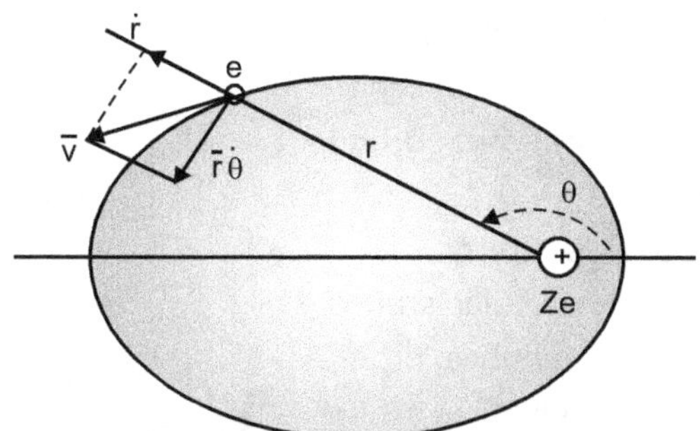

Fig. 1.8 : Sommerfeld's elliptical orbit

- Thus, in the case of elliptical orbit both the co-ordinates r and θ change with time. So we have to quantize the momenta associated with both these co-ordinates r and θ in accordance with Bohr's quantum condition.

- The two quantization conditions are

$$\oint P_r \, dr = rh \qquad \ldots (1.29)$$

and

$$\oint P_\theta \, d\theta = kh \qquad \ldots (1.30)$$

where r and k are radial and azimuthal quantum numbers respectively. Both r and k take only positive integral values.

- The detail analysis of Sommerfeld's theory gives the principal quantum number (n) as

$$n = r + k \qquad \ldots (1.31)$$

where

$$r = 0, 1, 2, 3 \ldots ,$$
$$k = 1, 2, 3 \ldots\ldots$$

and

$$n = 1, 2, 3 \ldots\ldots$$

- Thus, the principal quantum number n takes the integral values 1, 2, 3 …. Introduction of elliptical orbits do not generate new energy state for the hydrogen atom, but an electron may move in different orbits. Thus, for given principal quantum number n, the energy value is is fixed, but there are n different quantized orbits available for electron.

Examples :

For n = 1 : k = 1, r = 0, only one circular orbit is possible.

For n = 2 : k = 1, 2 and r = 0, 1.

Thus there are two possible orbits, these are

    k = 2 and r = 0   circular orbit

    k = 1 and r = 1   elliptical orbit.

For n = 3 : k = 1, 2, 3 and r = 0, 1, 2.

Thus there are three possible orbits.

k = 3 and r = 0	circular orbit
k = 2 and r = 1	elliptical orbit
k = 1 and r = 2	elliptical orbit

For n = 4 : k = 1, 2, 3, 4 and r = 0, 1, 2, 3.

Thus there exist four possible orbits.

k = 4 and r = 0	circular orbit
k = 3 and r = 1	elliptical orbit
k = 2 and r = 2	elliptical orbit
k = 1 and r = 3	elliptical orbit

- In general, if there are n number of orbits, then one is circular and remaining (n − 1) orbits are elliptical orbits. Fig. 1.9 shows the Sommerfeld elliptical orbits for n = 1, 2 and 3.

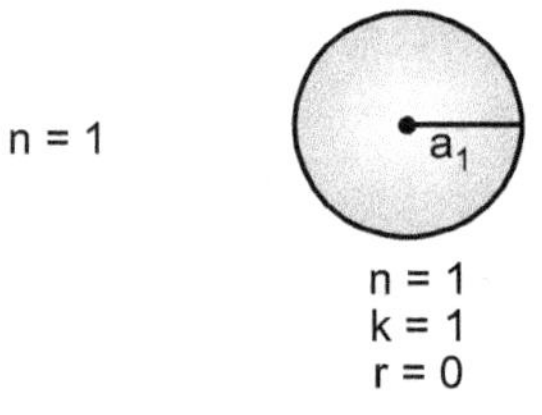

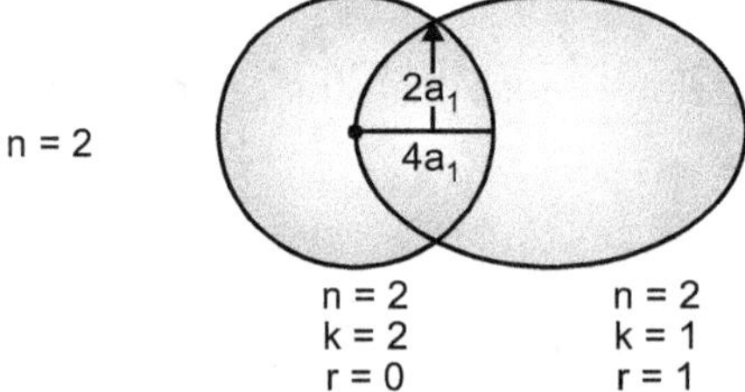

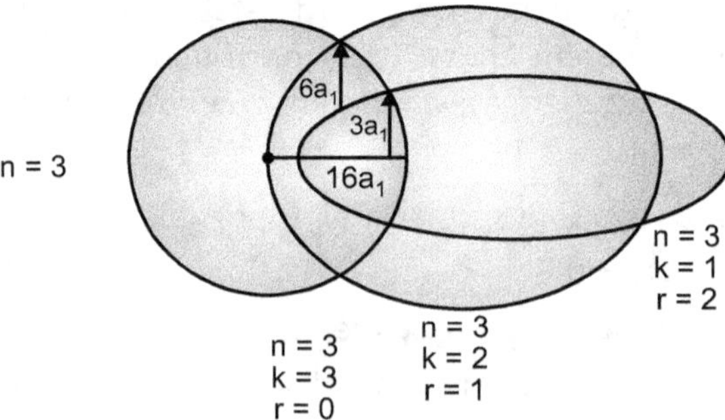

Fig. 1.9

- Each quantized orbit is determined by two out of the three quantum numbers n, k and r given by equation (1.31). Normally, total or principal quantum number n and azimuthal quantum number θ are used to specify the quantized state.

- According to this model, a given spectral line is produced by a number of different but equivalent transitions. For example, for n = 4, there are four possible orbits. All these orbits have same energy value as determined by total quantum number n. Such atomic states that can be described by more than one set of quantum numbers, which corresponds the same energy, are called **degenerate levels**. If the special theory of relativity is applied to the motion of electron, then in case of single electron system, degeneracy can be removed. The velocity of electron in an elliptical orbit is not constant. It is more when an electron is near the nucleus and less when it is far away from the nucleus.

- The mass of an electron varies with the velocity. According to the Sommerfeld, path of an electron is a precessing ellipse. The rate of precession is different for different orbits for the same **n value**. This gives n different states with slight difference in energies. This may remove the degeneracy and leads to the fine structure.

1.5 Vector Atom Model

- We have seen that Bohr's theory could not explain details of fine structure of spectral lines of hydrogen atom.

- Application of theory of relativity and the concept of elliptical orbit by Sommerfeld to the Bohr's theory, helped to explain the fine structure of the spectral lines of hydrogen atom. However, this theory could not predict the correct number of the fine structure lines and also does not give information about the relative intensities of the spectral lines. This theory could not explain the complex structure of alkali metals like sodium.

- Bohr and Sommerfeld's model could not explain new discoveries like Zeeman effect and Stark effect in which the spectral lines split up under the influence of magnetic and electric fields.

- The another drawback of Bohr model was that it could not explain how the orbital electron in an atom were distributed around the nucleus.

- Therefore, in order to explain the complex fine structure of spectra of atoms and their relation to atomic structure, a new model of an atom was suggested. This model is called vector atom model. The two distinct features of the vector atom model are :

(i)　The concept of space quantization.

(ii)　The concept of spinning electron.

Space Quantization :　　　　　　　　　　　　　　　　**(April 16, Oct. 15)**

- The space quantization means the electron orbits are quantized in space.

- As mentioned earlier, the motion of single electron of hydrogen like atom has been confined to two degrees of freedom. An electron orbit or energy state is determined by two of the three quantum numbers n, k and r. When the motion of an electron is considered in three-dimensional space, then this gives three degrees of freedom. This leads to necessity of three quantum numbers to specify each energy state, instead of two. Addition of third quantum number does not affect size and shape of the Bohr-Sommerfeld orbits, but it specifies orientation of orbits with reference to some direction in space. When an atom is placed in uniform magnetic field $\bar{H}$, then the fixed axis in space is considered along the direction of the magnetic field. Due to the magnetic field $\bar{H}$, the plane of the orbit precesses about $\bar{H}$, just like the top precesses in a gravitational field. The orbital angular momentum vector generates the cone about the field direction, $\bar{H}$. Fig. 1.10 shows the motion of electron with three degrees of freedom. The position of an electron in the polar coordinates is given by r, θ and ϕ. Reference axis is considered along $\bar{H}$. The angle δ is the angle between $\bar{H}$ and orbital angular momentum $\bar{L}$ of an electron. The vector $\bar{L}$ is normal to the orbit of an electron. Due to the quantum condition, the angle α takes certain discrete values. This gives rise to certain discrete orientations of orbits with respect to the $\bar{H}$. Thus the electron orbits are quantized in space. This is called **space quantization**. Consideration of three degrees of freedom leads to the three quantum numbers n, k and m. The quantum number m is called **magnetic quantum number** and takes the values

$$m = \pm1, \pm2, \pm3 \dots \pm k$$

where $\qquad k = 1, 2, 3, \dots n \qquad\qquad \dots (1.32)$

$$n = 1, 2, 3 \dots \infty$$

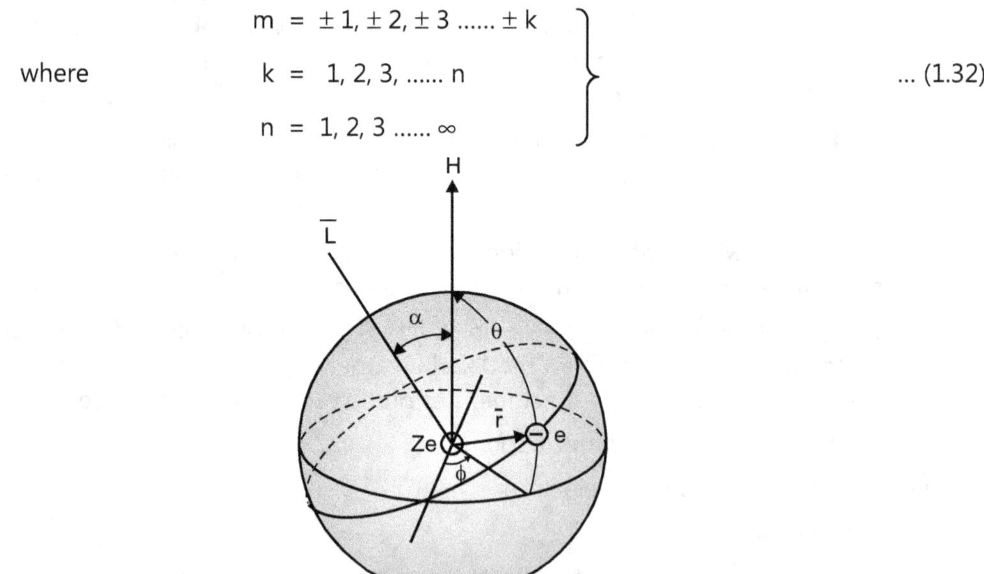

Fig. 1.10 : An electron motion in orbit with three degrees of freedom (r, θ and ϕ)

- In vector atom model, quantum numbers n, k and m given by Bohr-Sommerfeld's theory are replaced by new quantum numbers n, l and m_l as per the quantum mechanics. These are the principal quantum number (n), the orbital quantum number (l) and the orbital magnetic quantum number m_l. The values of these quantum numbers are

$$\left.\begin{array}{l} n = 1, 2, 3 \ldots\ldots\ldots \infty \\ l = 0, 1, 2, 3 \ldots\ldots (n-1) \\ m_l = -l, (-l+1), \ldots, 0, 1, \ldots (l-1), l \end{array}\right\} \qquad \ldots (1.33)$$

- According to the quantum mechanics, the orbital angular momentum of an electron is given by,

$$\bar{L} = \sqrt{l(l+1)} \left(\frac{h}{2\pi}\right) \qquad \ldots (1.34)$$

- The projection of $\bar{L}$ on the direction of magnetic field $\bar{H}$ (i.e. along the Z-axis) has magnitudes which are given by,

$$L_z = m_l \left(\frac{h}{2\pi}\right) \qquad \ldots (1.35)$$

where m_l is the orbital magnetic quantum number. Fig. 1.11 represents the space quantization of the orbital angular momentum for $l = 2$ and 3.

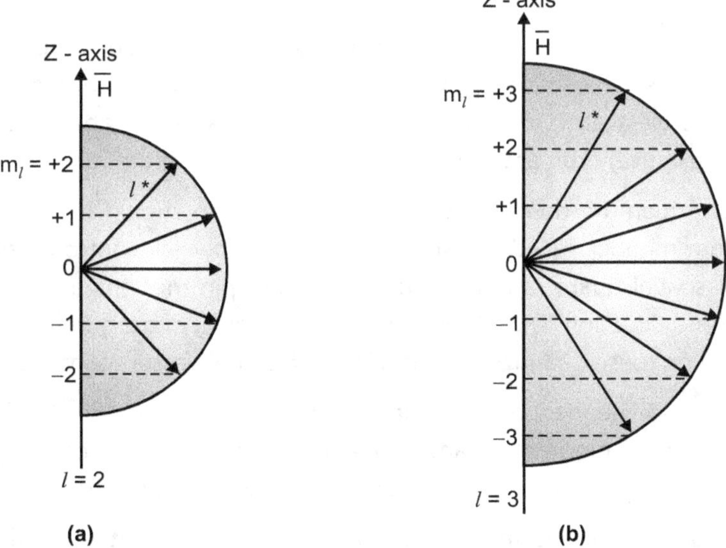

Fig. 1.11 : Space quantization of orbital angular momentum for $l = 2$ and $l = 3$

For $l = 2$, the number of orientations of $\bar{L}$ in space are $(2l + 1) = 5$. The magnitude of $\bar{L}$ is given by,

$$l^* = \sqrt{l(l+1)} \qquad \ldots (1.36)$$

$$\therefore \qquad l^* = \sqrt{2(2+1)} = \sqrt{6}$$

For $l = 3$, the number of orientations of $\bar{L}$ in space are

$$(2l + 1) \; = \; (2 \times 3 + 1) = 7$$

The magnitude of $\bar{L}$ is given by,

$$l^* \; = \; \sqrt{3\,(3 + 1)} = \sqrt{12}$$

The quantization of the direction of $\bar{L}$ with respect to the external magnetic field $\bar{H}$, is called space quantization.

Electron Spin :

- According to the quantum mechanics, the Schrodinger's equation, when applied to the hydrogen atom, results into three quantum numbers denoted by n, l, m_l for the specification of energy. These three quantum numbers could not explain unusual feature noted in the spectra of certain gases, such as sodium vapour. Close examination of one prominent line in the emission spectrum of sodium is a doublet. The wavelength of two lines of doublet are 5890 A° and 5896 A°.

- In order to solve this difficulty, S. Goudsmit and G. Uhlenbeck proposed the new concept of electron spin.

- According to their hypothesis, electron has spinning motion about an axis passing through its centre, while it also moves round the nucleus of an atom in its orbit. The spin motion of the electron is similar to the spinning of a planet about its own axis, as it moves in an elliptical orbit around the sun located at one of the focus.

- According to quantum theory, the spin of the electron should be quantized. Hence a new quantum number called the spin quantum number (s) is introduced. According to the idea of spatial quantization, orbital as well as spin motions are both quantized in magnitude and direction. Thus, they are considered as quantized vector. Hence the atom model based on these quantized vectors is called as the vector atom model.

- According to old theories, electron has supposed to have only orbital motion, hence only orbital angular momentum and orbital magnetic moment were considered. But, as electron also has spinning motion about an axis passing through its centre, an electron has spin angular momentum $(\bar{S})$ and spin magnetic moment (m_s). The spin angular momentum $(\bar{S})$ is given by

$$\bar{S} \; = \; \sqrt{s(s + 1)} \; \frac{h}{2\pi}$$

where, S is spin quantum number. Its value is $S = \dfrac{1}{2}$.

- In the presence of the magnetic field $\bar{H}$, the energy of an electron is slightly different for the two spin directions. This energy difference accounts for the existence of the sodium doublet. The quantum numbers corresponding to the spin of electron are

$$\left.\begin{array}{l} m_s = +\dfrac{1}{2} \text{ for the spin up state} \\[2ex] m_s = -\dfrac{1}{2} \text{ for the spin down state} \end{array}\right\} \qquad \text{... (1.37)}$$

- This is known as spin magnetic quantum number. According to the quantum mechanics, the physical spinning of an electron is not correct because the electron is a point particle, without spatial size. To overcome this conceptual difficulty, all the experimental proofs support that the electron does have some intrinsic property. This is called *spin of an electron* which results into the spin magnetic quantum number (m_s), the fourth quantum number. Sommerfeld and Paul Dirac pointed out that this quantum number has origin in the relativistic properties of an electron.

Quantum Numbers Associated with the Vector Atom Model   (Oct. 15, April 11)

- To describe the complete quantum state an electron in the atom, certain parameters are used. These sets of parameters are called quantum numbers. Schrodinger applied the quantum mechanics to the hydrogen atom and obtained the discrete energy states given by equation (1.44). The discrete nature of the energy states is the natural outcome of the Schrodinger's equation which is characterized by three quantum numbers- principal quantum number n, orbital angular momentum quantum number l, orbital magnetic quantum number m_l. The restrictions on these numbers are

$$n = 1, 2, 3, 4, \dots\dots$$
$$l = 0, 1, 2, 3 \dots\dots (n-1)$$
$$m_l = -l, (-l+1), \dots\dots 0, \dots\dots (l-1), l$$

- In addition to these three quantum numbers, the fourth quantum number called spin magnetic quantum number m_s is found to be necessary for the complete specification of the energy states. This spin magnetic quantum number came through the concept of electron spin which is intrinsic property of an electron as discussed in electron spin.

(a) **Principal quantum number (n) :** The *principal quantum number (n) specifies the orbit number, size of the orbit and the energy* of the state as given by equations (1.16) and (1.22) respectively.

$$r_n = \frac{\epsilon_0 h^2 n^2}{\pi m e^2}$$

$$E_n = -\left(\frac{m e^4}{8 \epsilon_0^2 h^2 n^2}\right)$$

- The above relations indicate that the size of orbit and energy state depends on principal quantum number.

$$r_n \propto n^2$$

$$E_n \propto \frac{1}{n^2}$$

The values of n are

$$n = 1, 2, 3, 4 \ldots\ldots \infty$$

which represent the energy levels or shells K, L, M, N, O, P …… respectively. The total number of electrons in a shell are $2n^2$. The serial number of shell starting from innermost is designated by n.

(b) Orbital quantum number (*l*) : This quantum number arises due to the orbital motion of an electron in the orbit. This gives rise to the orbital angular momentum of electron. On the basis of classical mechanics, it is given by,

$$\vec{L} = m\,(\vec{v} \times \vec{r})$$

where m is mass of electron, $\vec{v}$ is velocity and $\vec{r}$ the radius vector. $\vec{L}$ is perpendicular to the plane of electron orbit.

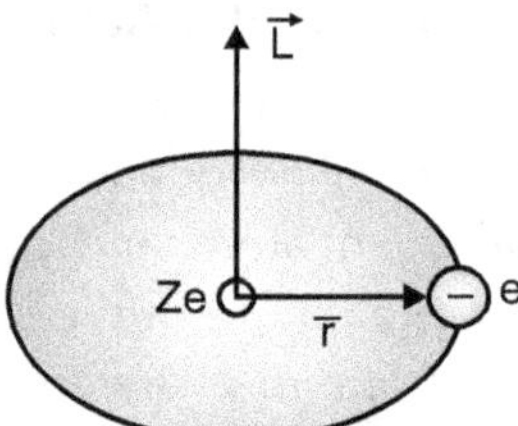

Fig. 1.12

- According to the Bohr's second postulate, orbital angular momentum $\vec{L}$ of electron is given by,

$$\vec{L} = n\frac{h}{2\pi}$$

- If L = 0 in Bohr's model, then it indicates that the motion of electron is along the straight line through the nucleus. This is not physically acceptable solution.

- The difficulties are resolved when the quantum mechanics is applied to the atomic system. According to the quantum mechanics, orbital angular momentum of an electron takes the discrete values as

$$\vec{L} = \sqrt{l\,(l + 1)}\,\frac{h}{2\pi} \qquad \ldots (1.37\ a)$$

where *l is called orbital angular momentum quantum number*. The value of *l* is restricted to

$$l = 0, 1, 2, 3 \ldots\ldots (n - 1)$$

n is the principal quantum number.

- Equation (1.37 a) shows that for L = 0, (l = 0) is an acceptable value for $\vec{L}$. Quantum mechanically, L = 0 indicates the spherically symmetric electron cloud with no axis of revolution.

- The values of l = 0, 1, 2, 3, 4, ... etc. represent the subshells s, p, d, f, g, ... etc. respectively.

  For example,

n = 1,	K-shell,	l has only one value,	l = 0	s-subshell
n = 2,	L-shell,	l has two values,	l = 0	s-subshell
			l = 1	p-subshell
n = 3,	M-shell,	l has three values,	l = 0	s-subshell
			l = 1	p-subshell
			l = 2	d-subshell

  and so on.

(c) Orbital magnetic quantum number (m_l) : If the atom is placed in weak magnetic field $\bar{H}$, then the orbital angular momentum $\vec{L}$ of an electron interacts with the applied magnetic field $\bar{H}$. According to the quantum mechanics, the quantization of the direction of $\vec{L}$ takes place in discrete manner. The component of $\vec{L}$ along the direction of external magnetic field along the Z-axis is given by

$$L_z = m_l \frac{h}{2\pi} \qquad \qquad \text{... (1.38)}$$

where m_l is called orbital magnetic angular momentum. The allowed values of m_l are

$$m_l = -l, (-l + 1), \ldots\ldots 0, \ldots\ldots (l - 1), l$$

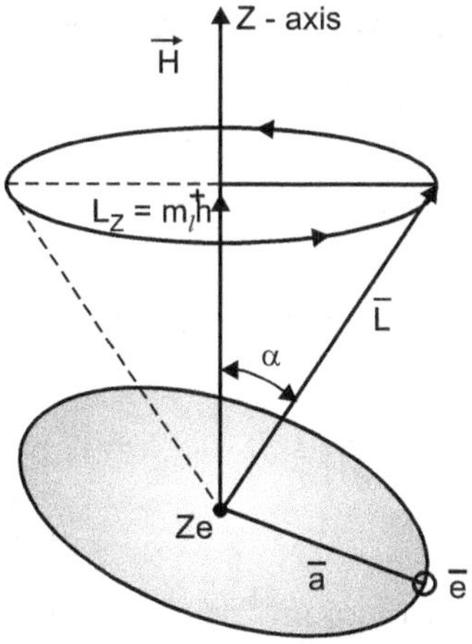

Fig. 1.13 : Vector $\vec{L}$ lies on the surface of a cone and precesses about $\vec{H}$

This shows that, for the given l, there are $(2l + 1)$ possible orientations of electron orbit in space with respect to the magnetic field direction (Z-axis). This is known as the space quantization which is discussed earlier in detail.

The angle between $\vec{L}$ and $\bar{L}_z$ is given by

$$\cos \alpha \ = \ \frac{L_z}{|\vec{L}|} = \frac{m_l}{\sqrt{l\,(l + 1)}}$$

For $l = 0$, $m_l = 0$ and $L_z = 0$

$l = 1$, $m_l = -1, 0, + 1$. This gives three discrete possible orientations of $\vec{L}$ in space.

$l = 2$, $m_l = -2, -1, 0, +1, +2$. This gives five possible orientations of $\vec{L}$ in space.

$l = 3$, $m_l = -3, -2, -1, 0, +1, +2, +3$.

This gives seven possible orientations of $\bar{L}$. The possible orientations of $\vec{L}$ are shown in Fig 1.13 for $l = 2$ and $l = 3$.

(d) Spin magnetic quantum number (m_s) : The spin magnetic quantum number (m_s) is due to the electron spin which is considered to be the intrinsic property of an electron. The spin angular momentum of an electron is given by equation

$$\bar{S} \ = \ \sqrt{s\,(s + 1)} \ \frac{h}{2\pi}$$

where s is spin quantum number. It's value is $\frac{1}{2}$. The spin angular momentum $\bar{S}$ is also

quantized, like $\bar{L}$, in space. It has two orientations with respect to the applied external

magnetic field $\bar{H}$. The component of $\bar{S}$ along the field direction (Z-axis) is given by,

$$S_z \ = \ m_s \frac{h}{2\pi} \qquad\qquad \text{... (1.39)}$$

where *m_s is called spin magnetic quantum number*. It has only two possible values $m_s = \pm\frac{1}{2}$.

The value $m_s = +\frac{1}{2}$ corresponds to spin-up, while $m_s = -\frac{1}{2}$ corresponds to the spin-down.

Fig. 1.14 represents the vector diagram for two allowed orientations of spin $\bar{S}$ in space with respect to the magnetic field direction $\bar{H}$ along Z-axis.

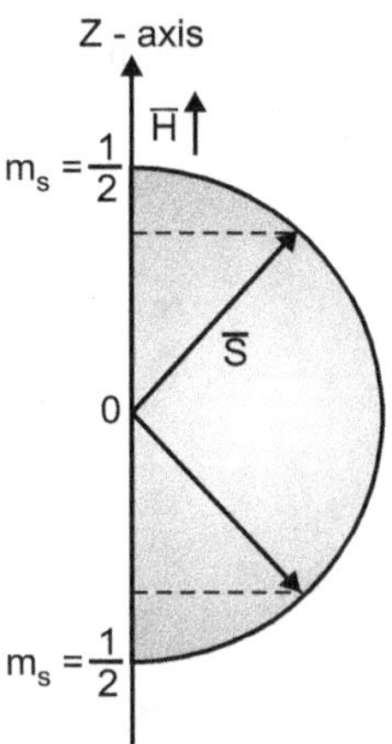

Fig. 1.14 : Two possible space orientations of $\overline{S}$

Total Angular Momentum (J) :

The *total angular momentum* of an electron is the vector sum of orbital angular momentum and spin angular momentum. Numerically it is expressed as

$$j = l \pm s \qquad \qquad ...(1.40)$$

But $s = \dfrac{1}{2}$ ∴ $\qquad j = l \pm \dfrac{1}{2} \qquad \qquad ...(1.41)$

According to quantum mechanics, total angular momentum of an electron is given by,

$$\overline{J} = \sqrt{j\,(j + 1)}\ \frac{h}{2\pi} \qquad \qquad ...(1.42)$$

where j is called total angular momentum of an electron.

1.6 Atomic Excitation and Atomic Spectra

- There are two methods in which an atom can be excited to an energy above its ground level and then become able to radiate. These are

1. Excitation by collision with another particle in which part of their joint K.E. is absorbed by an atom. Such excited atom comes to ground state in an average 10^{-8} s by emitting one or more photons.

   To produce discharge in rarefied gas, an electric field is established that accelerates electrons and atomic ions until their K.E. is sufficient to excite atoms they collide with. The energy transfer is maximum when colliding particle have same mass, the electrons in discharge are more effective than ions in providing energy to atomic electrons. Neon lamp and mercury vapour lamp are familiar examples of this. In this case, strong electric field is applied between electrodes in gas filled tubes which leads to emission of spectral radiation of gas (reddish light for Ne gas and bluish light in case of mercury vapour).

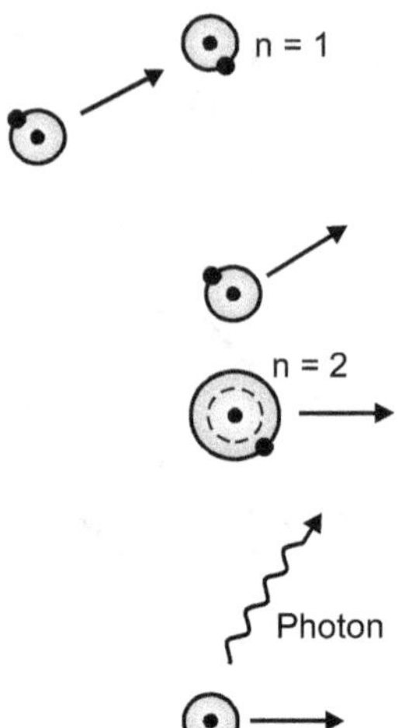

Fig. 1.15 : Excitation by collision

2. In another excitation mechanism atom absorbs a photon of light whose energy is just equal amount to raise the atom to a higher energy level. For example, photon of λ = 121.7 nm is emitted when an hydrogen atom in n = 2 state drops to n = 1 state. Absorption of photon of λ = 121.7 nm by a hydrogen atom initially in n = 1 state will bring it upto the n = 2 state. (Refer Fig. 1.16) This process explains origin of absorption spectra.

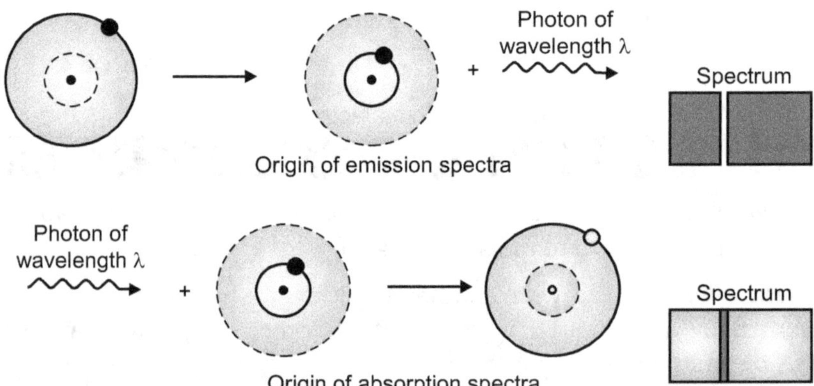

Fig. 1.16 : Origin of emission and absorption spectral lines

• When white light is passed through hydrogen gas, photons of those wavelengths that correspond to transition between energy levels are absorbed. The resulting excited hydrogen atoms reradiate their exciting energy at once, but these photons come off in random direction with only a few in same directions as the original beam of white light. (Refer Fig. 1.17)

- Hence dark lines in absorption spectra are never completely black, but only appear so by contrast with bright background.

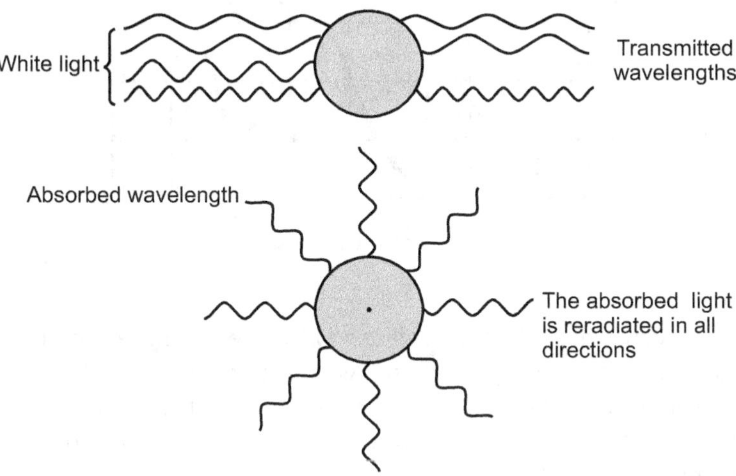

Fig. 1.17 : The dark lines in an absorption spectrum are never totally dark

FRANCK AND HERTZ EXPERIMENT

- A series of experiments based on excitation by collision was performed by Franck and Hertz in 1914. These experiments demonstrated that atomic levels exist and the atomic levels found in this way are same as those suggested by line spectra.

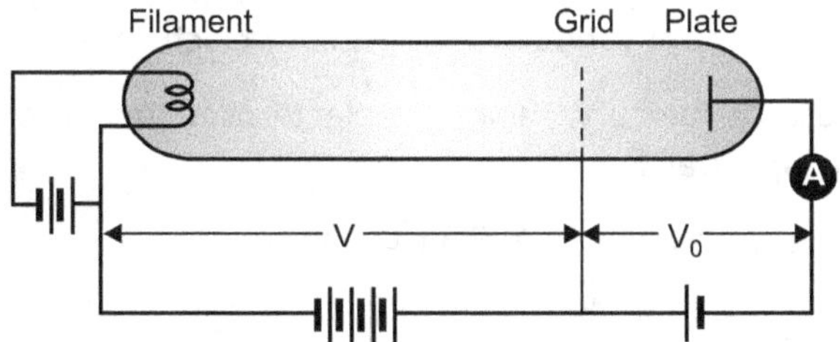

Fig. 1.18 : Apparatus for Franck-Hertz experiment

- Franck and Hertz bombarded the vapours of various elements with electron known energy using apparatus as shown in Fig. 1.18. A small potential V_0 is applied between grid and collecting plate preventing electron with less energy than certain minimum. As accelerating voltage V is increased, more and more electrons arrive at a plate and I rises. (Refer Fig. 1.19)

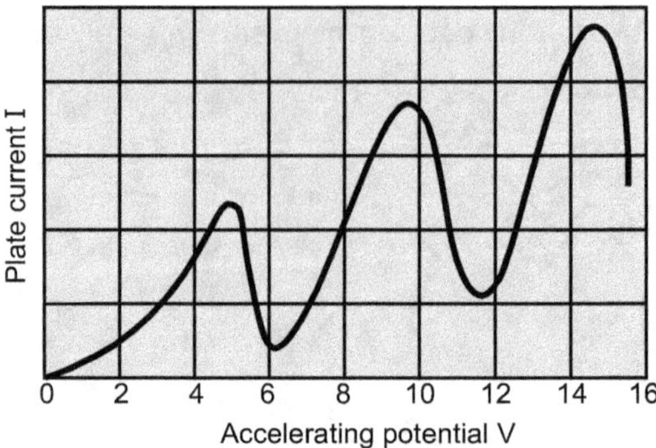

Fig. 1.19

- If K.E. is conserved when electron collides with one of the atoms in the vapour, the electron bounces off in new direction. As atoms are heavier than an electron, the electron loses almost no K.E. in the process. After certain critical energy is reached, the plate current drops abruptly. This suggests that an electron colliding with one of the atoms gives up some or all of its K.E. to excite the atom to an energy level above its ground state. Such collision is inelastic collision (in elastic collision, K.E. is conserved).

- Then as accelerating voltage is increased further, the plate current again increases since the electrons now have enough energy left to reach the plate current after undergoing inelastic collision. Another sharp drop in plate current occurs which arises from excitation of the same energy level in other atom by the electrons. Fig. 1.19 shows series of critical potentials for a given atomic vapour is obtained. Thus higher potentials result from two or more inelastic collisions and multiple of lowest one.

- To check whether the critical potentials are due to atomic energy levels, Franck or Hertz observed emission spectra of various vapour during electron bombardment.

- For example, in case of mercury, minimum electron energy of 4.9 eV was required to excite the 253.6 nm spectral line.

Theoretically photon of wavelength 253.6 nm has energy

$$E = h\upsilon$$

$$= \frac{hc}{\lambda}$$

$$= \frac{6.63 \times 10^{-34} \times 3 \times 10^{8}}{253.6 \times 10^{-9}}$$

$$= 7.84 \times 10^{-17} \text{ J}$$

$$= 4.9 \text{ eV}$$

Thus theoretical results agree with experimental observations as shown in Fig. 1.19.

Solved Problems

Problem 1.1 : *Experimentally it had been proved that energy of order of 13.6 eV is required to separate a hydrogen atom into proton and an electron. Find orbital radius and velocity of electron in a hydrogen atom.*

Solution : $E = -13.6 \text{ eV} = -13.6 \times 1.6 \times 10^{-19} = -2.2 \times 10^{-18} \text{ J}$

(a)
$$r = -\frac{e^2}{8\pi\varepsilon_o E} = -\frac{1.6 \times 10^{-19}}{8 \times 3.14 \times 8.85 \times 10^{-12}(-2.2 \times 10^{-18})}$$

$$= 5.3 \times 10^{-11} \text{ m} \qquad \text{... Ans.}$$

(b)
$$v = \frac{e}{\sqrt{4\pi\varepsilon_o\, mr}} = \frac{1.6 \times 10^{-19}}{\sqrt{4 \times 3.14 \times 8.85 \times 10^{-12} \times 9.1 \times 10^{-31} \times 5.3 \times 10^{-11}}}$$

$$= 2.2 \times 10^6 \text{ m/s} \qquad \text{... Ans.}$$

Problem 1.2 : *An electron collides with a hydrogen atom in its ground state and excites into state of n = 2. How much energy was given to hydrogen atom in this inelastic collision ?*

(April 16)

Solution : Energy change of an hydrogen atom that goes from its initial state of quantum number n_i to final state of quantum number n_f is

$$\Delta E = E_f - E_i = \frac{E_1}{n_f^2} - \frac{E_1}{n_i^2}$$

Here $n_i = 1$, $n_f = 2$ and $E_1 = -13.6$ eV

$$\Delta E = -13.6\left(\frac{1}{2^2} - \frac{1}{1^2}\right) = -13.6\left(\frac{1}{4} - \frac{1}{1}\right) = 10.2 \text{ eV} \qquad \text{... Ans.}$$

Problem 1.3 : *Calculate the linear velocity of an electron in the first, second and third orbit of hydrogen atom.*

(Oct. 17, 16)

Solution : The velocity of an electron in n^{th} orbit is given by,

$$v_n = \left(\frac{e^2 Z}{2\varepsilon_0 h}\right) \cdot \frac{1}{n}$$

For first orbit, we have $n = 1$, $Z = 1$.

$$v_1 = \frac{(1.6 \times 10^{-19})^2 \times 1}{2 \times (8.85 \times 10^{-12}) \times 6.64 \times 10^{-34}} \cdot \frac{1}{1}$$

$$v_1 = 2.17 \times 10^{-2} \times 10^{-38} \times 10^{46}$$

$$v_1 = 2.17 \times 10^6 \text{ m/s} \qquad \text{... Ans.}$$

For second orbit, $n = 2$.

$$\therefore \quad v_2 = 2.17 \times 10^6 \times \frac{1}{2}$$

$$\therefore \quad v_2 = 1.08 \times 10^6 \text{ m/s} \qquad \text{... Ans.}$$

For third orbit, $n = 3$.

$$\therefore \quad v_3 = 1.36 \times 10^6 \times \frac{1}{3}$$

$$\therefore \quad v_3 = 0.72 \times 10^6 \text{ m/s} \qquad \text{... Ans.}$$

Problem 1.4 : *Calculate the frequency of revolution of an electron in first and second orbit of hydrogen atom.*

Solution : The frequency of revolution of an electron in the nth orbit is given by

$$f_n = \left[\frac{me^4Z^2}{4\pi\,\varepsilon_0^2\,h^3}\right]\cdot\frac{1}{n^3}$$

For first orbit, n = 1; for hydrogen, Z = 1.

$$m = 9.11 \times 10^{-31}\,kg$$
$$e = 1.6 \times 10^{-19}\,C$$
$$\varepsilon_0 = 8.85 \times 10^{-12}\,C^2/Nm^2$$
$$h = 6.64 \times 10^{-34}\,Js$$

$$\therefore\qquad f_1 = \frac{9.11 \times 10^{-31}\times(1.6\times10^{-19})^4\times(1)^2}{4\times3.142\times(8.85\times10^{-12})^2\times(6.64\times10^{-34})^3}\times\frac{1}{(1)^3}$$

$$f_1 = \frac{59.6377\times10^{-107}}{2.8817\times10^5\times10^{-126}}$$

$$f_1 = 2.06\times10^{-4}\times10^{14}$$

$$f_1 = \mathbf{2.06\times10^{10}\ revolutions/s} \qquad\qquad \textbf{... Ans.}$$

For second orbit, n = 2.

$$f_2 = f_1\times\frac{1}{(2)^3}$$

$$f_2 = 2.06\times10^{10}\times\frac{1}{8}$$

$$f_2 = \mathbf{0.257\times10^{10}\ revolutions/s} \qquad\qquad \textbf{... Ans.}$$

Problem 1.5 : *Determine the magnitude of orbital angular momentum $\overline{L}$ and the allowed values of L_z and θ for the hydrogen atom in l = 3 states.*

Solution : The magnitude of orbital angular momentum $\overline{L}$ is given by,

$$|\overline{L}| = \sqrt{l(l+1)}\,\hbar$$

Since $l = 3$, $\therefore$

$$|\overline{L}| = \sqrt{3(3+1)}\,\hbar$$

$$|\overline{L}| = \sqrt{12}\,\hbar$$

The allowed values of L_z are given by,

$$L_z = m_l\hbar = m_e\left(\frac{h}{2\pi}\right)$$

For given l, m_e can have the following allowed values :

$$m_l = -l,\,(-l+1),\,0,\,(l+1),\,l$$

$$m_l = -3,\,-2,\,-1,\,0,\,1,\,2,\,3$$

$$\therefore\qquad L_z = -3\hbar,\,-2\hbar,\,-\hbar,\,0,\,\hbar,\,2\hbar,\,3\hbar$$

The allowed values of θ are

$$\cos\theta = \frac{L_z}{|\bar{L}|}$$

$$\cos\theta = \frac{m_l\,\hbar}{\sqrt{12}\,\hbar}$$

$$\theta = \cos^{-1}\left(\frac{m_l}{\sqrt{12}}\right)$$

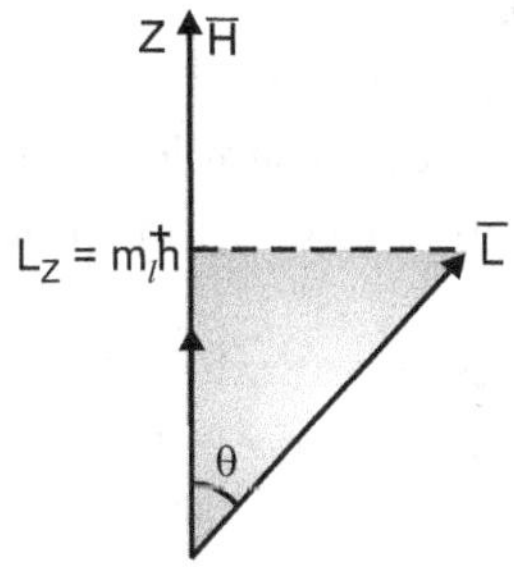

Fig 1.20

$\theta = 30.0°, 54.8°, 73.2°, 90.0°, 107°, 125°$ and $150°$.   ... **Ans.**

Summary

1. J.J. Thomson discovered electron and provided first model of an atom.
2. According to Rutherford's model, an atom is composed of tiny nucleus in which its positive charge and nearby all its mass are concentrated with electron some distance away to make atom electrically neutral as a whole.
3. Classically atomic electron spiral rapidly into nucleus as it radiates energy due to its acceleration.
4. Bohr modified the Rutherford's model by using the Planck's quantum theory of radiation. Bohr's theory is based on these postulates.
5. By Bohr's theory, energy of an electron is

$$E_n = -\frac{13.6}{n^2}\ eV$$

6. Balmer formula for wavelength of Balmer series is

$$\frac{1}{\lambda} = R\left(\frac{1}{2^2} - \frac{1}{n^2}\right),\ n = 3, 4, 5, \ldots\ldots$$

  where R = Rydberg constant = 1.097×10^7 m^{-1}

7. Vector atom model is based on the concept of space quantization and concept of spinning electron.
8. Space quantization : Electron orbits are quantized in space.
9. Spin of electron results into spin magnetic quantum number (m_s).

10. Principal quantum number (n), orbital quantum number (l), orbital magnetic quantum number and spin magnetic quantum number are major quantum numbers.

11. In excitation by collision some of the available energy is absorbed by one of the atoms, which goes into excited energy states. The atom then emits photon in returning to ground (normal) state.

12. Franck and Hertz experiment demonstrated that atomic levels exist.

Exercise

(A) Short Answer Type Questions :

1. Explain in short Rutherford's model of an atom.
2. What is the shortest wavelength present in Bracket series of spectral lines ?
3. In Bohr's model the electron is in constant motion. How can such electron have a negative amount of energy ?
4. What is the meaning of space quantization ?
5. What is the meaning of electron spin ?
6. State four quantum numbers.

(B) Long Answer Type Questions :

1. Find the expression for total energy and hydrogen atom in electron orbit.
2. Show that energy and electron in Bohr orbit is

$$E_n = -\frac{me^4}{8\varepsilon_o\, n^2\, h^2}$$

3. Discuss energy levels and series transition of hydrogen atom.
4. Discuss vector atom model.
5. What is atomic excitation ? Hence discuss origin of emission and absorption of spectral lines.
6. Discuss Franck and Hertz experiment. What conclusions are drawn from the experiment ?

(C) Numerical Problems :

1. An electron collides with a hydrogen atom in its ground state and excites to a state of n = 3. How much energy was given to hydrogen atom in this inelastic collision ? (**Ans.** 12.1 eV)
2. Find the largest wavelength present in the Balmer series of hydrogen corresponding to H_α line.
   (**Ans.** 656 nm)
3. Find frequency of revolution of electron in n = 1 and n = 2 Bohr orbit.
   (**Ans.** 6.58×10^{15} rev/s, 0.823×10^{15} rev/s.)
4. What is the frequency of the photon emitted when an electron in n = 2 orbit drops to n = 1 orbit ?
   (**Ans.** 2.88×10^{15} Hz)

Chapter **2**...
One Valence Electron Systems

Contents ...

Pauli Exclusion Principle

The **Pauli Exclusion Principle** is the quantum mechanical principle that states that two identical fermions (particles with half-integer spin) cannot occupy the same quantum state simultaneously. In the case of electrons, it can be stated as follows: it is impossible for two electrons of a poly-electron atom to have the same values of the four quantum numbers (n, l, m_l and ms). For two electrons residing in the same orbital, n, l, and m_l are the same, so m_s must be different and the electrons have opposite spins. This principle was formulated Austrian physicist Wolfgang Pauli in 1925.

Introduction

- In the earlier chapter, we have seen that the state of an electron in an atom is specified by four quantum numbers - principal quantum number n, orbital quantum number l, orbital magnetic quantum number m_l, and spin magnetic quantum number m_s. These four quantum numbers are used to describe all the possible electronic states, regardless of the number of electrons in an atom. In order to understand the structure of an atom, it is necessary to know the distribution of electrons into different shells and subshells. In principle, every electron in an atom prefers to enter into the lowest possible energy state of the atom. If this is true, then all electrons will enter in lowest possible energy state. As we know, such situation does not exist in nature. Answer to above question was given by the scientist W. Pauli in 1925, which is referred as the famous exclusion principle. In the present chapter, we will discuss the Pauli's exclusion principle, electron configuration, spectral notations of quantum states, spectra of single valence electron atom.

2.1 Pauli's Exclusion Principle

2.1.1 Statement   (April 17, 16; Oct. 16, 15)

- The Pauli's exclusion principle states that "*No two electrons in the same atom can have same four quantum numbers*". The principle may be stated as "*Every completely specified quantum state in an atom can be occupied by only one electron.*" Thus, exclusion principle puts certain restrictions on the values of four quantum numbers. When this principle is applied to the atom, the following two conclusions can be made. The first, for the given principal quantum number n, there can be no more than $2n^2$ electrons. Second, there can be only $2(2l + 1)$ electrons with the given value of l.

- For example, consider n = 1.

n	1	1
l	0	0
m_l	0	0
m_s	+1/2	−1/2

Thus, there are no more than two electrons in n = 1 state.

For n = 2, there are eight possibilities.

n	2	2	2	2	2	2	2	2
l	0	0	1	1	1	1	1	1
m_l	0	0	0	1	−1	0	1	−1
m_s	+1/2	−1/2	+1/2	+1/2	+1/2	−1/2	−1/2	−1/2

- Using Pauli's exclusion principle we can find the maximum number of electrons in a given group or subgroup. Let us calculate the maximum number of electrons in a given principal quantum number n. The calculation is based on the following rules:

(a) For a given n, the orbital quantum number l can have values 0, 1, 2, ...(n − 1).

(b) For particular value of l in (a), the azimuthal quantum number m_l has values −l, ... 0, ...+l i.e. in all $(2l + 1)$ values.

(c) For each m_l in (b), the spin quantum number m_s has two values namely, +1/2 and − 1/2 .

- The total number of states, for the given n (i.e. for shell) are

$$N = \sum_{l=0}^{n-1} 2(2l+1) \qquad \text{... (2.1)}$$

$$= 2[1 + 3 + 5 + \text{...} (2n-1)]$$

$$= 2\left[n\left\{\frac{1 + (2n-1)}{2}\right\}\right]$$

$$N = 2n^2 \qquad \text{... (2.2)}$$

- This is equal to the total number of electrons in the given shell.
- According to Pauli's principle, the third distinct set of quantum numbers is not possible for the third electron. Therefore for n = 1, only two electrons are possible (i.e. $2n^2$).

2.1.2 Electronic Configuration   (April 16, Oct. 15)

- *The electronic configuration of an atom is the distribution of electrons in various shells or subshells around the nucleus of the atom.* The electron orbits or shells in an atom are represented by the letters K, L, M, N, O, P, ... etc. corresponding to the principal or total quantum number n = 1, 2, 3, 4, 5, 6, 7, ... etc. respectively. The shells are divided into subshells. The electrons with same orbital quantum number l, form the subshell. The subshells are represented by small letters s, p, d, f, g, h, ... etc. corresponding to the orbital angular momentum quantum number l = 0, 1, 2, 3, 4, 5, 6, ... etc. respectively. These are represented in Table 2.1 below.

Table 2.1

Level	n	l	Sublevel notations	Number of electrons in sub-level = 2(2l + 1)	Total number of electrons in level = 2n²
K	1	0	1s	2	2
L	2	0	2s	2	8
		1	2p	6	
M	3	0	3s	2	18
		1	3p	6	
		2	3d	10	
N	4	0	4s	2	32
		1	4p	6	
		2	4d	10	
		3	4f	14	

- For $l = 0$, we get s-subshell; for $l = 1$, p-subshell; for $l = 2$, d-subshell; for $l = 3$, f-subshell and so on. The letters s, p, d, f have historical reference. The letter 's' indicates sharp series, p for principal series, d for diffuse and f for fundamental series. The state of electron is denoted by writing the value of principal quantum number n, before the correct letters which correspond to l - values.

For example,
$$1s \text{ state} : n = 1 \text{ and } l = 0$$
$$2s \text{ state} : n = 2 \text{ and } l = 0$$
$$2p \text{ state} : n = 2 \text{ and } l = 1$$
$$3s \text{ state} : n = 3 \text{ and } l = 0$$

The K-shell is the innermost one, which corresponds to n = 1.

For K-shell, $n = 1 \therefore l = 0$

The possible values of $m_l = 0$

The values of m_s are $m_s = +\dfrac{1}{2}$ and $-\dfrac{1}{2}$.

- With these values of n, l, m_l and m_s, we can have only two non-identical sets of quantum numbers as :

$$(n = 1, l = 0, m_l = 0, \; m_s = +\frac{1}{2})$$

$$(n = 1, l = 0, m_l = 0, \; m_s = -\frac{1}{2})$$

- No third set is possible. Therefore, according to the Pauli's exclusion principle, the K-shell can have at the most only two electrons, $(2n^2 = 2)$ corresponding to two distinct sets of quantum numbers.

- In the similar manner, we can proceed for L, M, N, O, P, ... etc. shells. In practice, we can understand the electronic structure of the complex atoms as succession of filled levels increasing in energy. In general, the order of filling of subshells in the atom is as follows : When a subshell is filled, the next electron goes into the lowest energy vacant sub-shell. Each subshell has particular number of orbitals. Each orbital has the capacity of only maximum two electrons. The state of electron in the given orbital is specified by three quantum numbers n, l and m_l. Out of two electrons in an orbital, one electron has a spin magnetic quantum number $m_s = +\dfrac{1}{2}$ and other electron has spin magnetic quantum number $m_s = -\dfrac{1}{2}$. The value $m_s = +\dfrac{1}{2}$ indicates that the spin vector of electron is pointed in upward direction, while $m_s = -\dfrac{1}{2}$ indicates that the spin vector is pointed in downward direction. The number of orbitals in given subshell are $(2l + 1)$. Thus, s-subshell $l = 0$, has one orbital; p-subshell $l = 1$, has three orbitals; d-subshell, $l = 2$, has five orbitals and so on. For the filling of electrons into subshells, Hund's rule is also obeyed in addition to the exclusion principle.

- The Hund's rule states that *"when an atom has orbital of equal energy, the order in which they are filled with electrons is such that a maximum number of electrons have unpaired spins."* Both, exclusion principle and Hund's rule can be illustrated through the electronic configuration of the following atoms.
- Let us consider the **hydrogen** atom, Z = 1.
- It has only one electron. This electron enters into 1s-subshell of K-shell. Electronic configuration is : 1 s^1
- The superscript 1 indicates that one electron is present in the 1s-subshell.
- Let us consider neutral **helium** atom, Z = 2.
- It has two electrons which enter into 1s-subshell of K-shell (n = 1). The electron configuration is 1s^2.

  For neutral **lithium**, Z = 3, there are 3 electrons.

  Electronic configuration : 1s^2 2s^1

  For **beryllium** atom, Z = 4, it has 4 electrons.

  Electronic configuration : 1s^2 2s^2

  For **boron** atom, Z = 5, it has 5 electrons.

  Electronic configuration : 1s^2 , 2s^2, 2p^1

  For **carbon** atom, Z = 6, it has 6 electrons.

  In this case, electronic configuration is 1s^2, 2s^2, 2p^2.

- Here two 2p-electrons cannot enter into only one orbital with paired spin (↑↓), but they enter into two separate orbitals with unpaired spins (↑↑). This is according to Hund's rule. The electronic configuration of the above elements is shown in Fig. 2.1.

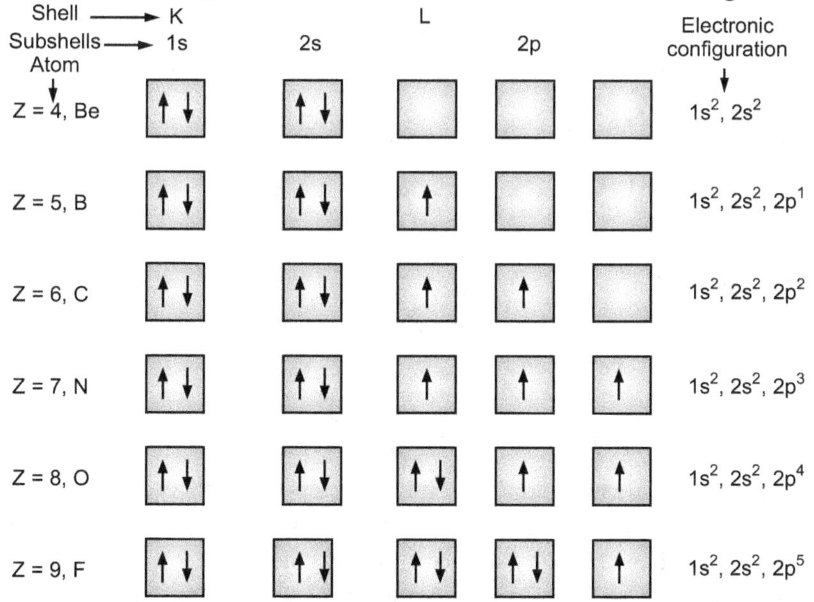

Fig. 2.1 : Electronic configuration for some lighter atoms

2.1.3 The Quantum State of an Electron   (April 16; Oct. 17, 16, 15)

- Every electron in an atom has its own unique set of four quantum numbers n, l, m_l and m_s. This set of quantum numbers specifies the state of an electron in an atom which is called *quantum state of an electron*. The total number of quantum states for the given l can be calculated as follows.

- Let us illustrate concept of quantum states by considering the example of **oxygen** atom.

- In oxygen atom, $Z = 8$, therefore, there are 8 electrons and electronic configuration is shown in Fig. 2.1.

First two electrons enter in s-subshell of K-shell ($n = 1$).

1st electron : $n = 1$, $l = 0$, $m_l = 0$, $m_s = +\dfrac{1}{2}$

2nd electron : $n = 1$, $l = 0$, $m_l = 0$, $m_s = -\dfrac{1}{2}$

Thus there are two non-identical quantum states for 1s-subshell.

Next two electrons enter in s-subshell of L-shell ($n = 2$).

3rd electron : $n = 2$, $l = 0$, $m_l = 0$, $m_s = +\dfrac{1}{2}$

4th electron : $n = 2$, $l = 0$, $m_l = 0$, $m_s = -\dfrac{1}{2}$

Thus there exist two quantum states for 2s-subshell.

The remaining four electrons enter into the 2p-subshell of L-shell.

For $l = 1$, m_l can have the values -1, 0, $+1$ and

For each m_l, m_s has two values $+\dfrac{1}{2}$ and $-\dfrac{1}{2}$.

5th electron : $n = 2$, $l = 1$, $m_l = -1$, $m_s = +\dfrac{1}{2}$

6th electron : $n = 2$, $l = 1$, $m_l = -1$, $m_s = -\dfrac{1}{2}$

7th electron : $n = 2$, $l = 1$, $m_l = 0$, $m_s = +\dfrac{1}{2}$

8th electron : $n = 2$, $l = 1$, $m_l = +1$, $m_s = +\dfrac{1}{2}$

- Though there are six allowed quantum states for 2p-subshell, but in case of oxygen, only 4-electrons are left after filling the 1s and 2s subshells, therefore only four quantum states exist. Thus, 2p-subshell is not completely filled with electrons. In oxygen atom, there exist 8 quantum states. This example also shows that for the completely filled subshell, orbital magnetic moment and spin magnetic moment becomes zero.

2.1.4 Spectral Notations of Quantum State　　　(April 17, 16; Oct. 17, 16, 15)

- For writing the atomic states, it is necessary to know the orbital angular momentum L, spin angular momentum S and the total angular momentum J of the atom. The total angular momentum J of an atom is the vector sum of orbital and spin angular momentum of an atom. It is given as

$$\vec{J} = \vec{L} + \vec{S} \qquad \qquad \text{... (2.3)}$$

where $\vec{L} = \sum \vec{l_i}$ is the orbital angular momentum of an atom. It is the vector addition of orbital angular momentum vector $\vec{l_i}$ of all the electrons.

$\vec{S} = \sum \vec{s_i}$ is the spin angular momentum of an atom. It is the vector sum of spin angular momentum $\vec{s_i}$ of all the electrons in the atom. The addition of the total orbital momentum of electron in atom $\vec{L}$ and total spin $\vec{S}$ gives total momentum $\vec{J}$. This type of coupling of $\vec{L}$ and $\vec{S}$ for determination of $\vec{J}$ is called L – S or Russell-Saunder's coupling.

- The total angular momentum vector $\vec{J}$ is characterized by the total angular momentum quantum number j. The total angular momentum $\vec{J}$ is also quantized. As per the quantum condition, $\vec{J}$ is quantized as

$$\vec{J} = \sqrt{j\,(j + 1)}\ \frac{h}{2\pi} \qquad \qquad \text{... (2.4)}$$

- The total angular momentum of an atom is decided by the last unpaired electrons in an atom. The core of electrons in an atom, do not contribute to the total angular momentum $\vec{J}$ of an atom. The total number of allowed values of $\vec{J}$ for the given $\vec{L}$ and $\vec{S}$ are

$$J = (L + S),\ (L + S - 1),\ ...\ (L - S + 1),\ |L - S| \qquad \text{... (2.5)}$$

For example, let L = 3 and S = $\frac{3}{2}$ then J = $\frac{9}{2}, \frac{7}{2}, \frac{5}{2}, \frac{3}{2}$

- Thus there are four values of $\vec{J}$. In general, there are (2S + 1) different values of $\vec{J}$. Each value of $\vec{J}$ gives a separate atomic state. This shows that there are (2S + 1) different energy states for the atom.

- The quantity (2S + 1) is called the multiplicity of the atomic state. The multiplicity (2S + 1) = 1 represents the singlet energy level (S = 0)

(2S + 1) = 2 represents the doublet energy levels (S = $\frac{1}{2}$).

(2S + 1) = 3 gives the triplet energy levels (S = 1).

It should be noted that, if L ≤ S, then the number of values of J are (2L + 1).

- The atomic states are denoted by the capital letters S, P, D, F, G, ... etc. which correspond to the L = 0, 1, 2, 3, 4, ... respectively. The atomic state is denoted using the term symbol as $^{(2S + 1)}L_J$, where $(2S + 1)$ is the multiplicity, $S = \sum s_i$ is the resultant spin momentum of atom, L is the resultant orbital angular momentum of atom.

$$L = 0, \qquad S \ - \ \text{state}$$
$$L = 1, \qquad P \ - \ \text{state}$$
$$L = 2, \qquad D \ - \ \text{state}$$
$$L = 3, \qquad F \ - \ \text{state and so on.}$$

Let us consider some examples.

Hydrogen atom in ground state :

$$n = 1, l = 0, m_l = 0 \text{ and } m_s = \frac{1}{2}$$

$\therefore$ State of an atom $^{(2S + 1)}L_J$ will be $^2S_{1/2}$

and read as doublet S half. In this notation, superscript 2 gives multiplicity symbol S for L = 0 and subscript 1/2 stands for J.

Similarly, L = 3 and S = 1/2

$$\text{The values of } J = \frac{7}{2}, \frac{5}{2}, \frac{3}{2} \text{ and } \frac{1}{2}$$
$$\text{The multiplicity} \Rightarrow (2S + 1) = 7$$
$$= \left(2 \times \frac{1}{2} + 1\right)$$
$$= 2$$

$\therefore$ L = 3 indicates F state as $^2F_{7/2,\ 5/2,\ 3/2,\ 1/2}$

- The atomic states : $^2F_{7/2}$, $^2F_{5/2}$, $^2F_{3/2}$ and $^2F_{1/2}$. All these states are doublets. The atomic state $^2F_{7/2}$ can be read as doublet F seven halves and so on.

- In general, the atomic level is described as $n^{(2S + 1)}L_J$, where n is principal quantum number.

2.2 Spectra of Single Valence Electron Systems

2.2.1 Spin-Orbit Interaction   (Oct. 15)

- The lines of the optical spectra emitted by alkali elements show a fine structure splitting which indicates that all energy levels such as P, D, F, G, etc. are doublet, except those for $l = 0$ (s level). This is due to spin-orbit interaction acting on the optically active electron, i.e. due to the coupling between the magnetic dipole moment of the electron and the internal magnetic field it feels because it moves through the electric field of the atom.

- Experimentally, it has been observed that any expression for doublet term separations will involve the atomic number Z, the quantum number l and total quantum number n. Calculation of the interaction energy due to the addition of an electron spin to the vector atom model, was carried out by Pauli, Darwin, Dirac, Gordon and others on the basis of quantum mechanics.

- *'The interaction energy is nothing but the change in kinetic energy of the system due to the precession of the electron around the field.*

- In the present section, we shall study the interaction between an electron's spin magnetic dipole moment and the internal magnetic field of a one-electron atom. Since the internal magnetic field is related to the electron's orbital angular momentum, this is called the *'spin-orbit interaction'*. It is a relatively weak interaction which is responsible, in part, for the fine structure of the excited states of one-electron atoms. A semiclassical calculation of the interaction energy may also be made.

- Consider an electron of mass m and charge −e moving with velocity v in a circular Bohr orbit of radius r (Refer Fig. 2.2). We may consider the motion of electron as a charge circulating in a loop. The circulating charge constitutes the current,

$$I = \frac{e}{T} \qquad \qquad ...(2.6)$$

where T is period of revolution. If v is the speed of electron then we have T = 2πr/v.

$$\therefore \qquad I = \frac{ev}{2\pi r}$$

For a current I in a loop of area A, the magnetic dipole moment is given by

$$\mu_l = IA = \frac{ev}{2\pi r} \times \pi r^2$$

$$\therefore \qquad \mu_l = \frac{evr}{2}$$

or $$\mu_l = \frac{e}{2m} mvr$$

The angular momentum is p_l = mvr.

$$\therefore \qquad \mu_l = \frac{e}{2m} p_l \qquad \qquad ...(2.7)$$

Since e is negative, its magnetic dipole moment μ_l is antiparallel to p_l as shown in Fig. 2.2. The quantity μ_l specifies the strength of the magnetic dipole: it equals the product of the poles' strengths times their separation. In CGS unit we write above equation as

$$\mu_l = \frac{e}{2mc} p_l \qquad \qquad ...(2.8)$$

where c is the speed of light.

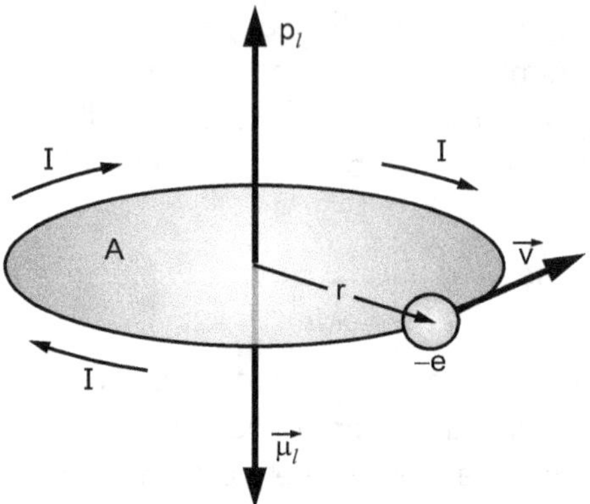

Fig. 2.2 : Orbital motion of electron and direction of current

Vectorially,

$$\vec{\mu_l} = -\frac{e}{2mc}\,\vec{p_l}$$

We know that

$$\frac{\mu_l}{p_l} = \frac{e}{2mc} \qquad \qquad \text{...(2.9)}$$

$$\therefore \quad \mu_l = p_l \cdot \frac{e}{2mc} = l\,\frac{eh}{4\pi mc} \qquad \left(\because p_l = l \cdot \frac{h}{2\pi}\right) \text{ ... (2.10)}$$

- The quantity $\dfrac{eh}{4\pi mc}$ in emu represents the smallest constant value of 'μ_l'. In terms of this smallest value, the magnitude of atomic and sub-atomic magnetic moments are measured. It is known as **Bohr magneton**. Its value is 9.21×10^{-21} erg-oersted^{-1} in emu or 9.27×10^{-24} joule $(Wb/m^2)^{-1}$ in MKS units.

- The motion of the magnetic moment of a particle, atom or molecule about the axis of an applied magnetic field is known as '*Larmour Precession*'. The magnetic moment sweeps out a cone about the direction of the magnetic field at a constant angular frequency as shown in Fig. 2.3.

- An electron behaves as if it possesses an intrinsic angular momentum $\dfrac{h}{2\pi}$ and magnetic moment. This electron moves in the electrical field $\vec{E}$ of the nucleus with velocity $\vec{v}$. The moving electron experiences a magnetic field given by,

$$\vec{H} = \frac{\vec{E} \times \vec{v}}{c^2} \qquad \qquad \text{... (2.11)}$$

According to the electromagnetic theory, the electric field due to nucleus is

$$\vec{E} = \frac{Ze}{r^3}\,\vec{r} \qquad \qquad \text{... (2.12)}$$

where 'Ze' is the charge on the nucleus and 'r' is the radial distance. From above two equations,

$$\vec{H} = \frac{Ze}{c^2 r^3}\, \vec{r} \times \vec{v}$$

or

$$\vec{H} = \frac{Ze}{mc^2 r^3}\, \vec{r} \times m\vec{v} \qquad \text{...(2.13)}$$

According to Bohr's quantum condition, the orbital angular momentum is

$$\vec{r} \times m\vec{v} = l^* \frac{h}{2\pi} \qquad \text{... (2.14)}$$

where, $l^* = \sqrt{l\,(l + 1)}$, m is the mass of the electron and v it's velocity.

Therefore, the field becomes

$$H = l^* \frac{h}{2\pi} \frac{Ze}{mc^2} \cdot \frac{1}{r^3} \qquad \text{... (2.15)}$$

- In this field the electron does not remain unaffected. The magnetic moment undergoes Larmor's precession around the filed direction as shown in Fig. 2.3. The angular velocity of the Larmour precession can be written as

$$\omega_L = \frac{H \cdot \mu_l}{p_l} \qquad \text{... (2.16)}$$

or

$$\omega_L = H \cdot \frac{e}{2mc} \qquad \text{... (2.17)}$$

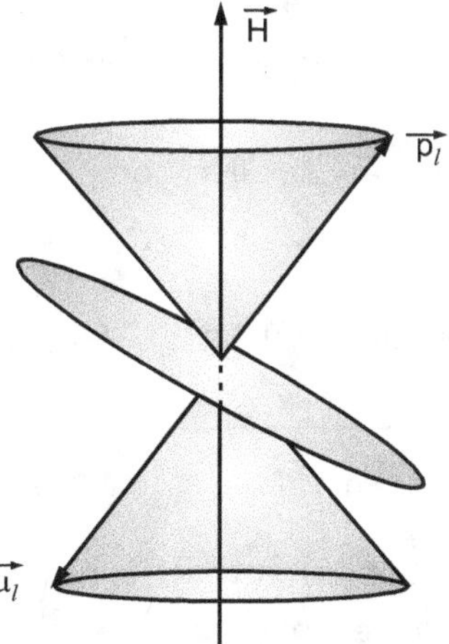

Fig. 2.3 : Precession of magnetic moment around the field direction

- If we take into account the spin of electron, the angular velocity is

$$\omega_L = H \cdot 2 \cdot \frac{e}{2mc}$$

- The factor '2' is due to the fact that the ratio for spin motion is double than that for orbital motion.

- If relativistic effect is considered, then there exists a relativistic precession of the electron in the opposite direction of the Larmor precession, the angular velocity of which is half that of Larmor type. Hence, the resultant precession is just half of that given by ω_L.

- Therefore, the resultant angular velocity,

$$\omega = \frac{1}{2}\omega_L = H \cdot \frac{e}{2mc}$$

$$\omega = \frac{1}{2} l^* \frac{h}{2\pi} \frac{Ze^2}{m^2c^3} \cdot \frac{1}{r^3} \qquad \text{... (2.18)}$$

- Now, the spin orbit interaction energy ΔW_{ls} is given as

$$\Delta W_{ls} = \begin{pmatrix} \text{Angular} \\ \text{velocity} \end{pmatrix} \times \begin{pmatrix} \text{Projection of spin} \\ \text{angular momentum} \\ \text{on } l^* \end{pmatrix}$$

$$\therefore \qquad \Delta W_{ls} = \omega \times s^* \frac{h}{2\pi} \cos(l^*s^*) \qquad \text{... (2.19)}$$

$$\therefore \qquad \Delta W_{ls} = \frac{1}{2} l^* \hbar \frac{Ze^2}{m^2c^3} \cdot \frac{1}{r^3} s^* \hbar \cos(l^*s^*) \qquad \left(\because \hbar = \frac{h}{2\pi} \right) \text{... (2.20)}$$

$$\therefore \qquad \Delta W_{ls} = \frac{Ze^2}{2m^2c^3} \hbar^2 \cdot \frac{1}{r^3} \cdot l^* s^* \cos(l^*s^*) \qquad \text{... (2.21)}$$

$$\therefore \qquad \Delta W_{ls} = \frac{Ze^2}{2m^2c^3} \frac{h^2}{4\pi^2} \cdot \frac{1}{r^3} \cdot l^*s^* \cos(l^*s^*) \qquad \text{... (2.22)}$$

- The radial distance 'r' is the function of Z, n and l and it changes continuously in any given state.

- Since the interaction energy is very small as compared to total energy, we need only to consider the average value of ΔW_{ls} and therefore of $\frac{1}{r^3}$.

- From perturbation theory and quantum mechanics,

$$\left< \frac{\overline{1}}{r^3} \right> = \frac{Z^3}{a_1^3 \, n^3 \, l \left(l + \frac{1}{2} \right)(l + 1)} \qquad \text{... (2.23)}$$

where 'a_1' is the radius of first Bohr orbit which is $\hbar^2/mc^2$.

- So the interaction energy, ΔW_{ls} becomes

$$\Delta W_{ls} = \frac{Ze^2}{2m^2c^3} \hbar^2 \cdot \frac{Z^2}{a_1^3 \, n^3 \, l \left(l + \frac{1}{2} \right)(l + 1)} \cdot l^* s^* (\cos l^* s^*) \qquad \text{...(2.24)}$$

For the factor $l^* s^* \cos(l^* s^*)$, we turn to the vector atom model.

- In a field free space, both l^* and s^* precess around their resultant j^*. According to law of conservation of momentum, the resultant j^* and hence the angle between l^* and s^* always remain invariant. It is illustrated in Fig. 2.4.

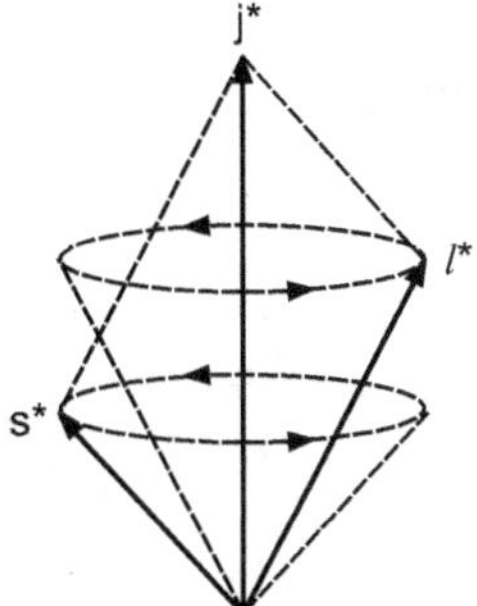

Fig. 2.4 : Precession of l^* and s^* around j^*

Using the Cosine law,

$$j^{*2} = l^{*2} + s^{*2} + 2l^*s^* \cos (l^*s^*) \qquad \text{... (2.25)}$$

$$\therefore \quad l^*s^* \cos (l^*s^*) = \frac{j^{*2} - l^{*2} - s^{*2}}{2} \qquad \text{...(2.26)}$$

Using equations (2.23) and (2.26), the equation (2.22) becomes,

$$\overline{\Delta W_{ls}} = \frac{Ze^2}{2m^2c^3} \hbar^2 \cdot \frac{Z^3}{a_1^3 \, n^3 l \left(l + \frac{1}{2} \right)(l + 1)} \cdot \frac{j^{*2} - l^{*2} - s^{*2}}{2} \qquad \text{... (2.27)}$$

$$\therefore \quad \overline{\Delta W_{ls}} = \frac{R\alpha^2 \, chZ^4}{n^3 l \left(l + \frac{1}{2} \right)(l + 1)} \cdot \frac{j^{*2} - l^{*2} - s^{*2}}{2} \qquad \text{... (2.28)}$$

where, $\qquad R = \dfrac{2\pi^2 \, me^4}{ch^3}$ is Rydberg constant

and $\qquad \alpha^2 = \dfrac{4\pi^2 \, me^4}{c^2 \, h^2}$ is the square of the fine structure constant.

This interaction energy is also called *term shift*.

Dividing equation (2.28) by hc, we get *term-shift in wave numbers* as,

$$\Delta\Gamma_{l,\,s} = \frac{- R\alpha^2 \, Z^4}{n^3 l \left(l + \frac{1}{2} \right)(l + 1)} \cdot \frac{j^{*2} - l^{*2} - s^{*2}}{2}$$

$$= -\Gamma \qquad \text{... (2.29)}$$

Negative sign indicates the convention that the term values are measured from series limit downwards.

This 'Γ' factor indicates spin-orbit interaction energy.

In short,
$$\Gamma = a\, l^*s^* \cos(l^*s^*)$$

$$\Gamma = a\,\frac{(j^{*2} - l^{*2} - s^{*2})}{2} \qquad \qquad ...(2.30)$$

where,
$$a = \frac{R\alpha^2 Z^4}{n^3\, l\left(l + \dfrac{1}{2}\right)(l + 1)}\ cm^{-1}$$

The term value Γ measures the separation from the hypothetical centre with term value Γ_0. Thus, the separation in term value due to spin orbit interaction is

$$\Gamma_j = \Gamma_0 - \Gamma$$

It is observed that

(a) The separation increases with increase in atomic number.

(b) The separation decreases with increase in principal quantum number 'n'.

(c) The separation decreases with increase in 'l'.

Following Fig. 2.5 shows term value separation for ^2P level.

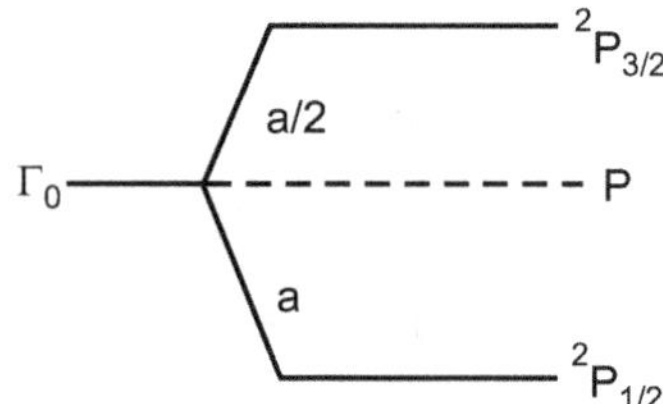

Fig. 2.5 : Splitting of P level due to spin-orbit interaction

2.2.2 Energy Levels of Sodium

- In order to incorporate the features like orbits with large 'n' values, higher series members and series limits into the same atomic picture, energy level diagrams are frequently drawn. In *energy level diagrams*, the vertical axis is an energy axis and the energy levels or terms are shown as horizontal lines. The difference between any two levels gives directly the frequency of the radiation in wave numbers in cm^{-1}.

- In atomic spectroscopy, the energy levels in the atom are described with a special name viz., *'terms'*. The energy value can be easily derived from the term value and also the representation by terms is much simpler. The zero of the energy is drawn at the top of the figure and other states below.

- Sodium has eleven electrons. The electron configuration of the normal sodium atom is represented by $1s^2\ 2s^2\ 2p^6\ 3s^1$. K and L shells are completely filled. The last level 3s is partially filled. The electron in this level plays a part in optical spectrum of sodium atom. The optical 3s electron of sodium atom when excited jumps to higher excited states 3p, 4s, 3d, 4p, 5s, 4d ...... etc. depending upon the amount of excitation energy. When this electron returns to ground level gives rise to spectral series. Depending on l values the levels are termed as S, P, D, F, G , ... etc. Since there is only one unpaired electron the spin is $s = \dfrac{1}{2}$ and multiplicity is $(2S+1) = 2$.

- The normal or ground state 3^2S of sodium arises from the electron configuration $1s^2\,2s^2\,2p^6\,3s^1$. Also the stationary states are designated as 3^2P, 3^2D ... etc. All substates belonging to a given n in sodium have not same value of energy because of shield of nuclear charge and penetration of atomic core. The 'l dependence' of energy is very important here, to the extent that 4S level lies lower than 3D orbit of the same orbit, similarly, 5S lies lower than 4D and 4F levels.

- This same designation scheme is used for the level assignments shown for sodium in Fig. 2.6. The electron configuration of each level is given by the excited electron only.

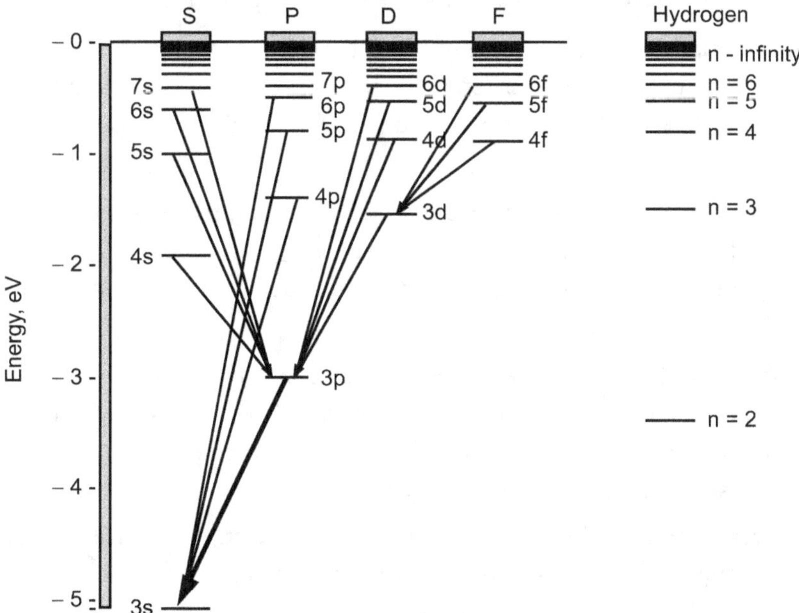

Fig. 2.6 : Energy levels of normal sodium atom and comparison with hydrogen

2.2.3 Selection Rules　　　　　　　　　　　　　　　　(April 16)

- General rules concerning the transitions, which may occur between the states of a quantum-mechanical, physical system, are known as *selection rules*. In other words, the transition rules in spectroscopy are known as selection rules. Physically, the selection rules arise because of symmetry properties of the oscillating charge distribution of the atom. It is experimentally found that all the possible combinations of permitted energy states of an atom don't actually appear as spectral lines.

- We have seen that transitions can take place between 'S' and 'P' states, 'P' and 'D' states, 'D' and 'F' states, but under normal conditions no transitions can take place between 'S' and 'D' states, 'S' and 'F' states or 'P' and 'F' states. It is because of certain restrictions in transitions. These restrictions are called 'Selection Rules'.

- On the basis of Selection Rules, transitions are called 'allowed' or 'forbidden'. These rules were first introduced on a purely empirical basis, but they were obtained by a rigorously deductive method in wave mechanics and hence they now rest on a permanent theoretical basis. These selection rules have been devised for L, for J and for S etc.

- The selection rules are as follows :

(a) The selection rule for L : For this quantum number, the rule is $\Delta L = \pm 1$, i.e., transitions take place only between levels whose L quantum number differs by one i.e. only these lines are observed for which the value of L changes by ± 1.

(b) The selection rule for J : It is $\Delta J = \pm 1$ or 0. ($0 \leftrightarrow 0$ excluded). $\Delta J = 0$ is also permissible because the direction of spin of an electron in an orbit can have only two possible values $\pm 1/2$, so that a change in 'S' of ± 1, when added vectorially to $\Delta L = \pm 1$ can lead to $\Delta J = 0$.

(c) The selection rule for S : For this quantum number, the rule is $\Delta S = 0$ which means that states with different 'S' (hence different multiplicities) do not combine with one another. Theory and experiment, however, show that this selection rule is only an approximate rule holding good in the case of light atoms.

(d) The selection rule for total quantum number (n) : In Bohr's theory, there was no restriction on changes in 'n' for transitions of atoms from one energy level to another. Similarly, according to quantum theory, there is no restriction on changes in 'n'. It can change by any positive integral rule, including zero.

(e) The selection rule for quantum number l or n_θ : The rule is, $\Delta l = \pm 1$ and $\Delta n_\theta = \pm 1$. (i.e. in any electron transition it must change by +1 or −1).

(f) The selection rule for m_l and m_s : The orbital magnetic quantum number, m_l either does not change or changes by ± 1, i.e. $\Delta m_l = 0$ or ± 1.

The spin magnetic quantum number 'm_s' remains unchanged, i.e. $\Delta m_s = 0$.

In consequence, $\Delta m_j = 0$ or ± 1.

These selection rules are statements of transition probabilities that are based on quantum mechanical calculations. These rules are useful in allotting proper quantum numbers to observe spectral lines in the series. With their aid, energy level diagrams can be constructed for natural complex multiplet lines and Zeeman effect lines.

2.2.4 Spectra of Sodium Atom　　　　　　　　　　　　　　(April 17, 16)

Sodium is one of the alkali metals. The optical spectrum of sodium is typical spectrum of all the alkali atoms. Since there is only one unpaired electron, the spin is $s = \frac{1}{2}$ and multiplicity is $(2S+1) = 2$. The spectral terms are depicted in the following table

Sub level	L	J	Spectral terms
S	0	1/2	$^2S_{1/2}$
P	1	1/2, 3/2	$^2P_{1/2}, {}^2P_{3/2}$
D	2	3/2, 5/2	$^2D_{3/2}, {}^2D_{5/2}$
F	3	5/2, 7/2	$^2F_{5/2}, 2F_{7/2}$

- The energy level diagram of sodium is shown in Fig. 2.7. It shows different lines in different series and their transitions in the spectrum of sodium. According to selection rules certain transitions are allowed and listed below. These transitions of lines can be represented in four main series as follows :

(1) Sharp series :

$$\text{Transitions} - n\,{}^2S_{1/2} \rightarrow 3\,{}^2P_{3/2,\,1/2}\,,\ \text{doublet}$$

$$n = 4, 5, 6 \dots$$

(2) Principal series :

$$\text{Transitions} - n\,{}^2P_{3/2,\,1/2} \rightarrow 3\,{}^2S_{1/2},\ \text{doublet}$$

$$n = 3, 4, 5 \dots$$

(3) Diffuse series :

$$\text{Transitions} - n\,{}^2D_{5/2,\,3/2} \rightarrow 3\,{}^2P_{3/2,\,1/2,}\ \text{triplets}$$

$$n = 3, 4, 5 \dots$$

(4) Fundamental series :

$$\text{Transitions} - n\,{}^2F_{7/2,\,5/2} \rightarrow 3\,{}^2D_{5/2,\,3/2,}\ \text{triplets}$$

$$n = 4, 5, 6 \dots$$

- As sodium source contains a large number of atoms, all possible excited states are available for transition, hence all the series are simultaneously observed. Note that in Fig. 2.7, the $^2P_{3/2}$ level is actually slightly above the $^2P_{1/2}$ level as j value is greater. The separation of levels decreases as the value of n (total quantum number) increases.

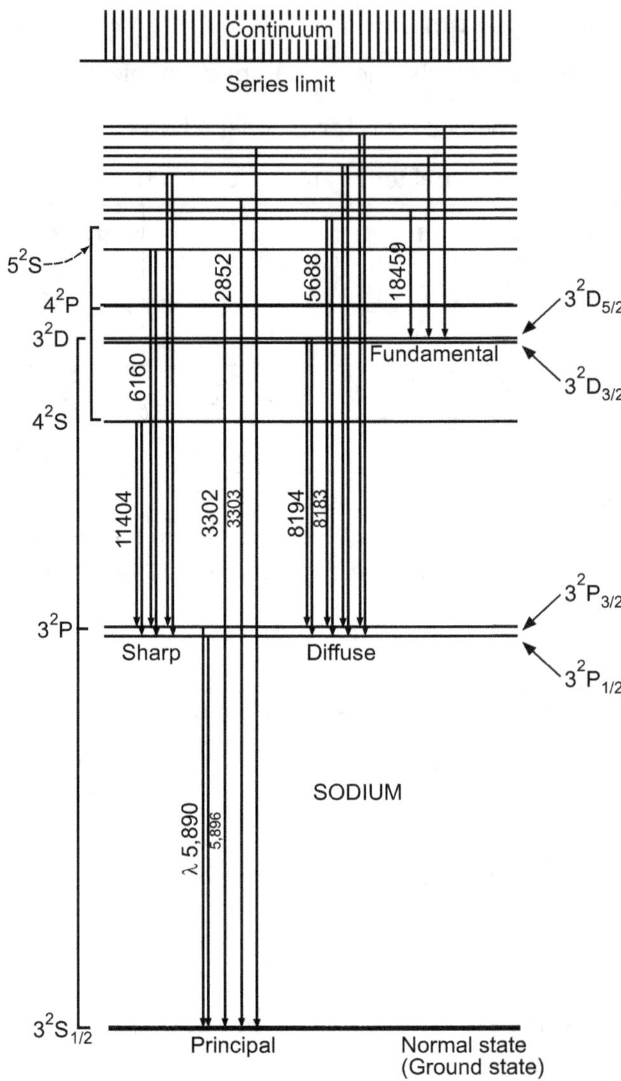

Fig. 2.7 : Transitions of spectral lines in energy level diagram of sodium atom

2.2.5 The Sodium Doublet

- The classical example is the yellow sodium doublet lines which are also known as sodium 'D' lines at 589.592 nm and 588.995 nm. The D_1 component (5896 A°) is due to the transition $P_{1/2} \rightarrow S_{1/2}$ and the 'D$_2$' component (5890 A°) to $P_{3/2} \rightarrow S_{1/2}$ as shown in Fig. 2.8. This well-known bright doublet which is responsible for the bright yellow light from a sodium lamp may be used to demonstrate several of the influences which cause splitting of the emission lines of atomic spectra.

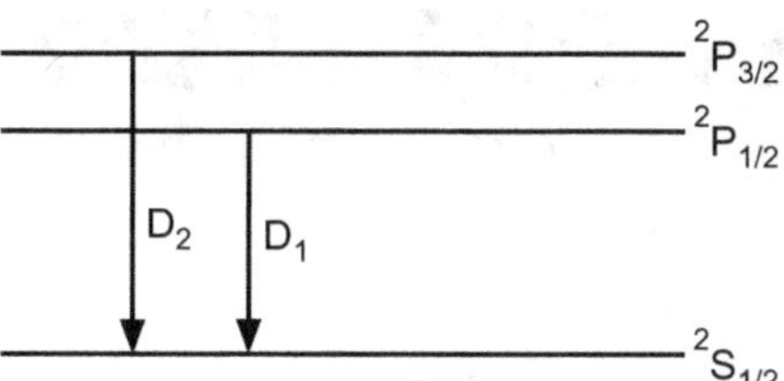

Fig. 2.8 : Principal series doublet

- Applying the selection rules $\Delta L = \pm 1$ and $\Delta J = 0$ or ± 1 (excluding $0 \rightarrow 0$), both the transitions are allowed, which explain the doublet fine structure of the sodium 'D' line. The diffuse and fundamental series start from doublet levels and end on doublet levels but the selection rules only permit triplets.

- The doublet character of Na-spectrum is shown in Fig. 2.9.

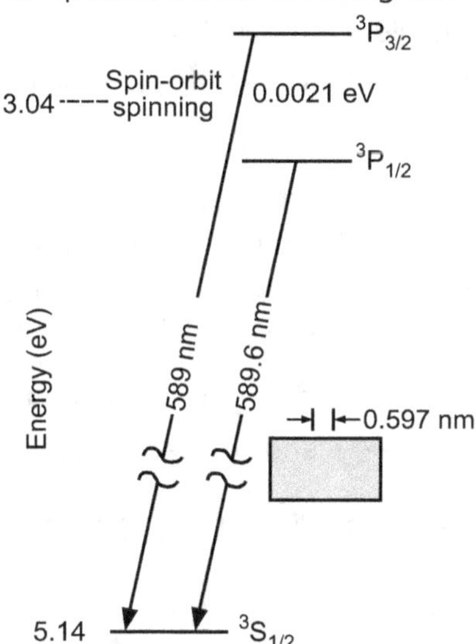

Fig. 2.9 : Sodium doublet

Some points are also noted in regard to alkali spectra.

(1) Corresponding doublet separation increases with atomic number.

(2) Doublet separations in the ionised alkaline earths are larger than the corresponding doublets in the alkali metals.

(3) Within each element, doublet separations decrease in going to higher members of a series.

(4) Within each element 'P' doublets are wider than 'D' doublets of the same 'n' and 'D' are wide than 'F' of the same 'n'. Fig. 2.9 explains these facts. Hence, in alkali spectra every component level is double excluding 'S' level because for 'S' state $l = 0$, so the possible value of j is only one i.e. 1/2.

Solved Problems

Problem 2.1 : *Write the electronic configuration of fluorine and neon.*

Solution : For fluorine F, Z = 9.

$\therefore$ There are 9 electrons in fluorine atom.

The electronic configuration will be $1s^2$, $2s^2$, $2p^5$.

For Ne atom, Z = 10.

$\therefore$ There are 10 electrons in Ne atom.

The electronic configuration will be $1s^2$, $2s^2$, $2p^6$.   **... Ans.**

Problem 2.2 : *Determine the ground state of aluminium atom. Represent it using the spectral notations.*   **(Oct. 17, 16)**

Solution : Consider the aluminium atom Al, Z = 13.

There are 13 electrons in Al atom. The shells and subshells are filled as follows :

$$1s^2,\ 2s^2\ 2p^6,\ 3s^2\ 3p^1$$

The electrons in 1s, 2s, 2p and 3s subshells are paired and are completely filled. The last 13th electron enters into 3p - subshell. This electron is unpaired. The paired electrons in the inner subshells do not contribute to the orbital and spin angular momentum of the atom. Therefore, the last unpaired electron in 3p-subshell decides the ground state and the total angular momentum $\overline{J}$ of the whole atom. Thus last unpaired electron : 3p-electron state.

$\therefore$ $n = 3$, $l = 1$, $m_l = -1, 0, +1$ and for each value of m_l, there are two values of $m_s = \pm \dfrac{1}{2}$. Since only one electron is available in 3p-subshell, its state is specified by the quantum numbers $n = 3$, $l = 1$, $m_l = -1$ and $m_s = +\dfrac{1}{2}$.

Thus, $L = l = 1$

$$S = s = \frac{1}{2}$$

$\therefore$ Multiplicity $= (2S + 1) = 2$

$$\text{Values of } J = \frac{3}{2}, \frac{1}{2}$$

Thus, the ground state of Al atom can be expressed as $^2P_{3/2,\ 1/2}$.   **... Ans.**

Problem 2.3 : *If L = 2, S = $\dfrac{1}{2}$, then write the atomic states.*   **(Oct. 16)**

Solution : Given : $L = 2$, $S = \dfrac{1}{2}$

$$\text{The values of } J = \frac{5}{2}, \frac{3}{2}$$

$$\text{Multiplicity} = (2S + 1) = \left(2 \times \frac{1}{2} + 1\right) = 2$$

$\therefore$ $L = 2$ represents D-state.

Thus, the atomic states can be expressed as $^2D_{5/2}$ and $^2D_{3/2}$.   **... Ans.**

Problem 2.4 : *Determine the total number of quantum states of electrons in carbon atom.*

Solution : For carbon atom, $Z = 6$, therefore, there are six electrons.

First two electrons enter in 1s-subshell of K-shell ($n = 1$).

1st electron : ($n = 1$, $l = 0$, $m_l = 0$, $m_s = +\frac{1}{2}$)

2nd electron : ($n = 1$, $l = 0$, $m_l = 0$, $m_s = -\frac{1}{2}$)

Next two electrons enter in 2s-subshell of L-shell ($n = 2$).

3rd electron : ($n = 2$, $l = 0$, $m_l = 0$, $m_s = +\frac{1}{2}$)

4th electron : ($n = 2$, $l = 0$, $m_l = 0$, $m_s = -\frac{1}{2}$)

Last two electrons enter in 2p-subshell of L-shell ($n = 2$).

5th electron : ($n = 2$, $l = 1$, $m_l = -1$, $m_s = +\frac{1}{2}$)

6th electron : ($n = 2$, $l = 1$, $m_l = -1$, $m_s = -\frac{1}{2}$)

Thus there are six quantum states of electrons in carbon atom.   **... Ans.**

Problem 2.5 : *Find the S, L and J value that corresponds to each of the following :*

$$^1S_0, \; ^3P_2, \; ^3D_{3/2}, \; ^2P_2$$

Solution : For $^1S_0$:

$$L = 0, \; (2S + 1) = 1$$

$\therefore$ $S = 0$ and hence $J = 0$

For $^3P_2$:

$$L = 1, \; (2S + 1) = 3$$

$\therefore$ $S = 1$ and hence $J = 2$ or zero

[As $J = 2$ is admissible value with $L = 1$ and $S = 1$, $^3P_2$ represents an allowed state with $J = 2$]

For $^2D_{3/2}$:

$$L = 2, (2S + 1) = 2 \therefore S = \frac{1}{2}. \text{ Hence } J = \frac{5}{2} \text{ and } \frac{3}{2}$$

As $J = \frac{3}{2}$ is an admissible value with $L = \frac{1}{2}$ and $S = \frac{1}{2}$, $^2D_{3/2}$ represents value with allowed term.

For $^2P_2$:

$$L = 1, (2S + 1) = 2 \therefore S = \frac{1}{2}. \text{ Hence } J = \frac{3}{2} \text{ and } \frac{1}{2}$$

As $J = 2$ is not admissible value with $L = \frac{1}{2}$ and $S = \frac{1}{2}$, $^2P_2$ does not represent an allowed state.	**... Ans.**

Problem 2.6 : *The frequency of strong yellow line in the spectrum of sodium is $5.09 \times 10^{14} sec^{-1}$. Calculate the wavelength of the light in nanometers.*

Solution : We have the relation,

$$\lambda = \frac{c}{\upsilon}$$

Here, in this case,	$c = 3.0 \times 10^8 \, m \, sec^{-1}$

$$\upsilon = 5.09 \times 10^{14} \, sec^{-1}$$

$\therefore$	The wavelength of light,

$$\lambda = \frac{3 \times 10^8}{5.09 \times 10^{14}} = 589 \times 10^{-9}$$

$\therefore$	$\lambda = \textbf{589 nm}$

Thus the wavelength of sodium light is 589 nm.	**... Ans.**

Problem 2.7 : *Find the orbital angular momentum of 'd' electron.*

Solution : For 'd' electron, $l = 2$.

As the orbital angular momentum is given as,

$$l^* = \sqrt{l \, (l + 1)} \, \hbar$$

$\therefore$	$$l^* = \sqrt{l \, (l + 1)} \, \frac{h}{2\pi}$$

$$l^* = \frac{\sqrt{2 \, (2 + 1)} \times 6.63 \times 10^{-34}}{2 \times 3.14}$$

$\therefore$	$l^* = \textbf{2.58} \times \textbf{10}^{-34} \, \textbf{Js.}$	**... Ans.**

Thus, the orbital angular momentum for 'd' electron is 2.58×10^{-34} Js.

Problem 2.8 : *Calculate the value of electronic angular momentum of a one-electron atom in the state $^2D_{5/2}$.*

Solution : The given term is $^2D_{5/2}$.

For this term,

$$\text{Multiplicity } m = 2S + 1 = 2, \ S = \frac{1}{2} \text{ and } L = 2.$$

$$\therefore \qquad J = L + S = 2 + \frac{1}{2} = \frac{5}{2}$$

The total electronic angular momentum is given by,

$$J^* = \sqrt{J(J+1)}\,\frac{h}{2\pi}$$

$$\therefore \qquad J^* = \sqrt{\frac{5}{2}\left(\frac{5}{2}+1\right)} \times \frac{6.63 \times 10^{-34}}{2 \times 3.14}$$

$$\therefore \qquad J^* = 3.123 \times 10^{-34} = \mathbf{3.123 \times 10^{-34}\ Js} \qquad\qquad \textbf{... Ans.}$$

Problem 2.9 : *The ionisation energy of sodium is 494.7 kJ mol^{-1}. Calculate the wavelength of the electromagnetic radiation which is just sufficient to ionise the sodium atom.*

(Given : $c = 3 \times 10^8\ ms^{-1}$; $h = 6.6 \times 10^{-34}\ Js$; $N_A = 6.02 \times 10^{23}$).

Solution : The ionisation energy of sodium is

$$\Delta E = 494.7 \text{ kJ/mol}$$

$$= \frac{494.7 \times 10^3}{6.02 \times 10^{-23}} \text{ J/atom}$$

$$\therefore \qquad \lambda = \frac{hc}{\Delta E}$$

$$= \frac{6.6 \times 10^{-34} \times 3 \times 10^8 \times 6.02 \times 10^{23}}{494.7 \times 10^3}$$

$$\lambda = \mathbf{24.09 \times 10^{-7}\ m} \qquad\qquad \textbf{... Ans.}$$

Problem 2.10 : *If the doublet splitting of the first excited state $2\ ^2P_{3/2} - 2\ ^2P_{1/2}$ of He$^+$ is 5.84 cm^{-1}, calculate the corresponding separations of H.*

Solution : The doublet splitting of a one-electron atomic state arising due to spin-orbit interaction is given by,

$$\Delta T = \frac{R\,\alpha^2\,Z^4}{n^3 l\left(l + \dfrac{1}{2}\right)(l + 1)}\ \text{cm}^{-1}$$

where   R is Rydberg constant,

        α is fine structure constant and

        Z is atomic number

$\because$ For a given state, (n, l constant), therefore we have,

$$\Delta T \propto Z^4$$

For He$^+$ and H, Z = 2 and 1 respectively.

$\therefore$
$$\frac{\Delta T_{He^+}}{\Delta T_H} = \frac{(2)^4}{(1)^4} = 16$$

$\therefore \quad \Delta T_H = \dfrac{1}{16}$ $\therefore \Delta T_{He^+} = \dfrac{1}{16} \times 5.84 \text{ cm}^{-1}$

$\therefore \qquad \qquad \Delta T_H = \mathbf{0.365 \text{ cm}^{-1}}$ **... Ans.**

Summary

1. **Pauli's exclusion principle :** No two electrons in the same atom can have same quantum numbers.

2. **Electronic configuration** of an atom is the distribution of electrons in various shells or subshells around nucleus of the atom.

3. **Hund's rule :** When an atom has orbital of equal energy, the order in which they are filled with electrons is such that a maximum number of electrons have unpaired spins.

4. Every electron in an atom has its unique set of quantum numbers n, l, m_l, m_s. This set of quantum numbers specifies the state of an electron in an atom which is called quantum state of electron.

5. Electron states are denoted by small letters s, p, d, f, g, h, ... corresponding to the orbital angular quantum number l = 0, 1, 2, 3, 4, ... respectively.

6. For writing atomic states one should know L and S so that J = L + S, where J is total angular momentum. The atomic states are denoted by capital letters S, P, D, E, F, ... corresponding to L = 0, 1, 2, 3, 4, ... respectively. The atomic state is denoted using term symbol as $^{(2S + 1)}L_J$, where (2S + 1) is multiplicity.

7. **Spin-orbit interaction** is a weak interaction which is responsible for the fine structure of excited states of one electron atom.

8. The motion of the magnetic moment of a particle, atom or molecule about axis of applied magnetic field is known as Larmour Precession.

9. The quantity $\dfrac{eh}{4\pi mc}$ in emu represents the smallest constant value of μ_l called Bohr magneton.

$$\frac{eh}{4\pi mc} = 9.21 \times 10^{-27} \text{ erg·oersted}^{-1}$$

$$= 9.27 \times 10^{-24} \text{ J}$$

10. **Interaction energy (or term shift)** in spin-orbit interaction is

$$\Delta\Gamma_{l,\,s} = \frac{-R\alpha^2 Z^4}{n^3\, l\,(l + 1/2)\,(l + 1)}\; \frac{j^{*2} - l^{*2} - s^{*2}}{2}$$

11. In energy level diagram, the vertical axis is an energy axis and the energy levels or terms are shown as horizontal lines.

12. General rules concerning the transition which may occur between the states of quantum mechanical, physical system are known as selection rule. These are

$$\Delta L = \pm 1$$
$$\Delta J = \pm 1 \text{ or } 0$$
$$\Delta S = 0$$
$$\Delta l = \pm 1$$
$$\Delta m_l = 0 \text{ or } \pm 1$$
$$\Delta m_j = 0 \text{ or } \pm 1$$

13. Spectra of sodium atom consists of sharp series, principal series, diffused series and fundamental series.

14. In alkali spectra, corresponding doublet separation increases with atomic number.

Exercise

(A) Short Answer Type Questions :

1. State the Pauli's exclusion principle.
2. What is an electronic configuration of sodium ?
3. What is the orbital angular and spin angular momentum of completely filled subshells ?
4. What is the limit for the number of subshells in a shell ?
5. What is the quantum state of an electron ?
6. What is multiplicity of the state ?
7. How do you calculate the values of $\overline{J}$?
8. Define spin-orbit interaction energy.
9. State Larmour theorem.
10. What is spin-orbit interaction ?
11. Why 4s level is filled first by electrons than 3d level in outer atomic shells of atom ?
12. What is the selection rule for the orbital quantum number for electric dipole transitions in a single electron atom ?
13. What is the S value for $^2D_{3/2}$?
14. What is the 'L' value for state $^2D_{3/2}$?
15. What are the values of total angular momentum of single 'f' electrons ?
16. What are selection rules ?
17. What are the selection rules for quantum numbers 'm_s' and 'm_l' ?
18. What are the selection rules for quantum numbers 'L' and 'S' ?
19. What are the 'L' and 'S' quantum numbers corresponding to $^2D_{3/2}$?

(B) Long Answer Type Questions :

1. State and explain the Pauli's exclusion principle.
2. Explain the electronic configuration with suitable examples.
3. Discuss the quantum states of an electron.
4. Explain the representation of an electron state and the atomic state using the spectral notations.
5. What is multiplicity ? Explain its importance in spectroscopy.
6. Show that the total number of quantum states for the given n are $2n^2$.
7. Derive spin-orbit interaction energy expression. What is the significance of negative sign ?
8. Draw energy level diagram and discuss the energy levels of sodium.
9. What are selection rules ? Describe selection rules in connection with different quantum numbers.
10. Mention uses of selection rules.
11. Discuss the spectrum of sodium, explaining fine structure of sodium D lines.
12. Give the representation of spectral lines of different series in sodium spectrum in terms of symbols.
13. Derive the expression for ω_L for spin-orbit interaction. What is the change in ω_L due to relativistic treatment ?
14. Describe the main features of the doublet character of sodium spectrum.
15. Discuss the spectra of sodium atom and draw
    (a) Energy level diagram. (b) Fine structure.

(C) Numerical Problems :

1. Determine the ground state of sodium atom. Represent it using the spectral notations.
2. Determine the total number of quantum states of electrons in magnesium atom.
3. Calculate the atomic states for

    (a) $L = 4$ and $S = \dfrac{1}{2}$ (b) $L = 3$ and $S = 1$

    (c) $L = 2$ and $S = \dfrac{3}{2}$ (d) $L = 5$ and $S = \dfrac{3}{2}$.

4. Determine the ground state of cadmium (Cd) atom (Z = 48). Represent it using the spectroscopic notations.
5. If one of the states of the configuration is $^6H_{6/2}$, what are the other possible states ?
6. Calculate the values of (a) l, s and j (b) $L, S,$ and J
   for a 'd' electron in one-electron system.
7. Use the vector model of the atom to determine possible values of the total angular momentum for a sodium, when principal quantum number n = 3. Draw an energy level diagram and indicate the transitions you would expect to occur. Calculate the angles between $\vec{l}$ and $\vec{s}$ vectors in each case.

Chapter **3**...
Two Valence Electron Systems

Contents ...

(13 December 1888–30 October 1976) was a German-American physicist known for his contributions to quantum theory. He is responsible for the Landé g-factor and an explanation of the Zeeman Effect. In atomic physics, the **Landé interval rule** states that if the spin-orbit interactions of an electron are weak, the energy levels of each (i.e. the spin and orbit) are split. Subsequently, each have a different angular momentum. The rule states that as a result of this, the interval between successive energy levels is proportional to the larger of their total angular momentum values

**German American Physicist
- Lande**

Introduction

•   We have already seen in the vector atom model that each component part is assigned a quantum number, the numerical value of which may conveniently be thought of as the length of the vector which represents the angular momentum of that component. In last chapter, we have studied the atomic spectrum having single valence electron. The alkaline-earths such as beryllium, magnesium, calcium etc. contain two valence electrons outside the closed shells or subshells. On excitation of atoms of these elements either one or both the electrons are excited and give rise to spectra of these elements.

- In this chapter, we will study the two valence electron atoms and the coupling schemes involved between angular momenta of two electrons. We will also study the interaction energies in such systems.

3.1 Spectral Terms of Two-Electron Atom

- The alkaline-earths have two valence electrons. Since an atom consists of two valence electrons with different orbital and spin angular momenta, a coupling scheme is necessary to obtain resultant angular momentum. Each electron is assigned a spin angular momentum $s^* \dfrac{h}{2\pi}$ where $s^* = \sqrt{s(s+1)}$ and $s = \dfrac{1}{2}$ in addition to the orbital angular momentum $l^* \dfrac{h}{2\pi}$, where $l^* = \sqrt{l(l+1)}$ ($l = 0, 1, 2, 3, ...$). Two-electron system is one of the many-electron system and in these systems each electron has its 'l' and 's' or 'l^*' and 's^*' vectors. The vector addition of all angular momentum vectors would give the total angular momentum (J) of the atom.

- There are several ways in which different vectors of the electrons may combine to give the vectors representing the atom as a whole. The method of combination depends on the interaction (or coupling) between the component vectors. The coupling schemes are of two types: (1) LS coupling and (2) jj coupling.

3.1.1 LS Coupling   (Oct. 16)

- This coupling scheme is also called *Russell Saunders coupling*. In this type of coupling, several $\vec{l}$ vectors combine to form resultant $\vec{L}$ and $\vec{s}$ vectors form resultant $\vec{S}$.

$$\vec{L} = \vec{l_1} + \vec{l_2} + \vec{l_3} + ...$$

$$\vec{S} = \vec{s_1} + \vec{s_2} + \vec{s_3} + ...$$

- The resultant $\vec{L}$ and $\vec{S}$ then combine to form resultant $\vec{J}$ which represent the total angular momentum i.e.

$$\vec{J} = \vec{L} + \vec{S}$$

- Consider the orbital motions of two electrons and let l_1 and l_2 represent their respective orbital quantum numbers. Their angular momenta are $l_1^* \dfrac{h}{2\pi}$ and $l_2^* \dfrac{h}{2\pi}$ or simply l_1^* and l_2^* respectively.

- On quantum mechanical approach, l_1^* and l_2^* are quantized with respect to each other in such a way that they form a resultant $L^* = l_1^* + l_2^*$

where $L^* = \sqrt{L(L+1)}$ and L takes the values from $|l_1 - l_2|$ to $|l_1 + l_2|$

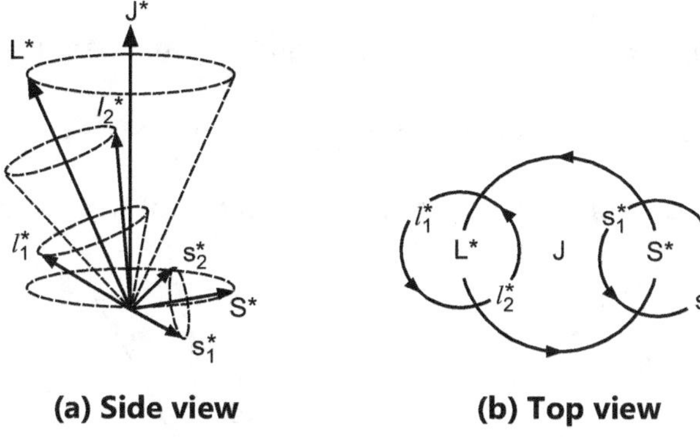

(a) Side view   **(b) Top view**

Fig. 3.1 : LS coupling of two electrons

Fig. 3.2 shows L values for $l_1 = 1$ and $l_2 = 1$. These values give L = 0, 1, 2.

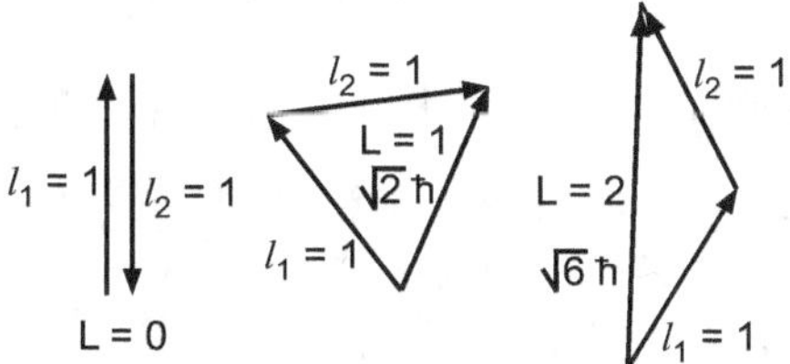

Fig. 3.2 : Schematic diagram suggesting how $l_1 = 1$ and $l_2 = 1$ give L = 0, 1 or 2

For $l_1 = 1$ and $l_2 = 1$, $|l_1 - l_2| = 0$, and $|l_1 + l_2| = 2$, therefore, L = 0, 1, 2.

For L = 0, $L^* = 0$.

For L = 1, $L^* = \sqrt{1(1+1)}\,\dfrac{h}{2\pi} = \sqrt{2}\,\dfrac{h}{2\pi}$

For L = 2, $L^* = \sqrt{2(2+1)}\,\dfrac{h}{2\pi} = \sqrt{6}\,\dfrac{h}{2\pi}$

- Fig. 3.3 shows L values for $l_1 = 1$ and $l_2 = 2$. These values give L = 1, 2, 3.

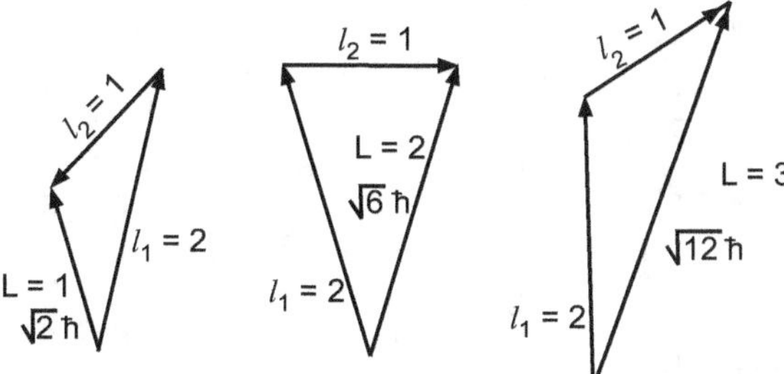

Fig. 3.3 : Schematic diagram suggesting how $l_1 = 1$ and $l_2 = 2$ give L = 1, 2 or 3

- With two electrons, each having a spin angular momentum of $s^* \dfrac{h}{2\pi}$ where $s^* = \sqrt{s(s+1)}$ and $s = \dfrac{1}{2}$, there are two ways in which a spin resultant $S^* \dfrac{h}{2\pi}$ may be formed. Let s_1^* and s_2^* represent the respective spin vectors of two electrons.

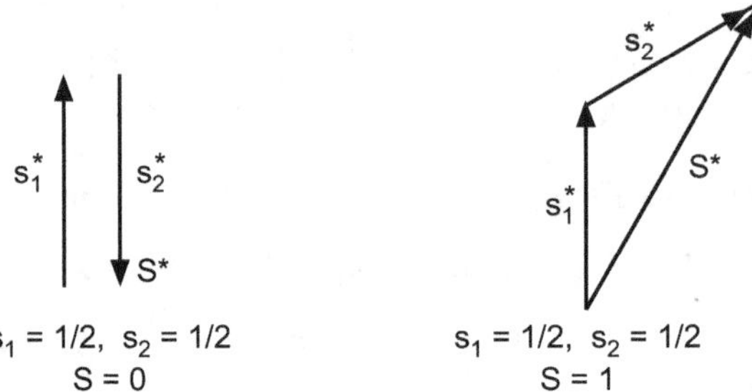

Fig. 3.4 : Vector diagrams of ss coupling of two-valence electrons

- Quantizing these (as shown in Fig. 3.4) we get, $s_1^* = \dfrac{1}{2}\sqrt{3}$ and $s_2^* = \dfrac{1}{2}\sqrt{3}$. The two resultants are one with $S^* = 0$ and the other with $S^* = \sqrt{2}$. Thus, the resultant quantum values are $S = 0$ and $S = 1$. The resultant $S^* = 0$ will give rise to singlet terms while $S^* = \sqrt{2}$ give rise to triplet terms.

- Finally L^* and S^* after coupling form J^* which represents the total angular momentum of the atom $J^* \dfrac{h}{2\pi}$. The guiding principles in this type of coupling show that J must always be positive, never negative and $J^* = \sqrt{J(J+1)}$ and J takes values as $|L - S|, \ldots, |L + S|$.

- Electron configurations commonly occurring in two-electron systems are s·s, p·d, p·f, d·d, d·f, etc. Consider as a specific example, the case of one electron in a 'p' orbit and the other in a 'd' orbit. i.e. two electrons in a 'pd' configuration in LS coupling.

  For 'p' electron,

$$l_1 = 1 \text{ and } s_1 = \frac{1}{2}$$

Similarly, for 'd' electron,

$$l_2 = 2 \text{ and } s_2 = \frac{1}{2}$$

∴　L values in this case are,

$$L = (2+1), \ldots (2-1) = 3, 2, 1$$

and
$$S = \text{'0' or '1'}.$$

The multiplicity of the state is given by,

$$m = (2S + 1).$$

Here
$$m = 1 \text{ or } 3.$$

- Therefore, the state is either singlet or triplet. Thus, two-electron system gives rise to singlet states for $S = 0$ and $m = 1$ and triplet states when $S = 1$ and $m = 3$. For singlets or triplets with $L = 1, 2, 3$ corresponding spectral terms are 'P', 'D', 'F'.

- Also, J has values such that $J = |L - S|, |L - S| + 1, \ldots, L + S - 1, L + S$.

- Therefore the term symbols for $S = 0$ and $J = L = 1, 2, 3$ are $^1P_1$, $^1D_2$, $^1F_3$ which correspond to three singlet terms.

  With $S = 1$ there are three possibilities for each of $L = 1, 2, 3$ are

  (i) When $S = 1$, $L = 1$ ('P' terms) then 'J' has values lying between $(1 + 1)$ and $(1 - 1)$ i.e. 0, 1, 2 values. The spectral terms are $^3P_0$, $^3P_1$, $^3P_2$ or can be written as $^3P_{0,1,2}$.

  (ii) When $S = 1$, $L = 2$ ('D' terms) then 'J' has values lying between $|2 - 1|$ and $|2 + 1|$ i.e. 1, 2, 3 values. The spectral terms are $^3D_1$, $^3D_2$, $^3D_3$ or can be written as $^3D_{1,2,3}$.

  (iii) When $S = 1$, $L = 3$ ('F' terms) then 'J' has values lying between $|3 - 1|$ and $|3 + 1|$ i.e. 2, 3, 4 values. The spectral terms are $^3F_2$, $^3F_3$, $^3F_4$ or can be written as $^3F_{2,3,4}$.

- The terms S, P, D, F, ... give the value of L, and the subscript gives the resultant value of J. The superscript to left expresses multiplicity.

- Hence, 'pd' configuration of two electrons in LS coupling gives in all twelve terms, three singlet and nine triplet as shown in the vector diagram in Fig. 3.5.

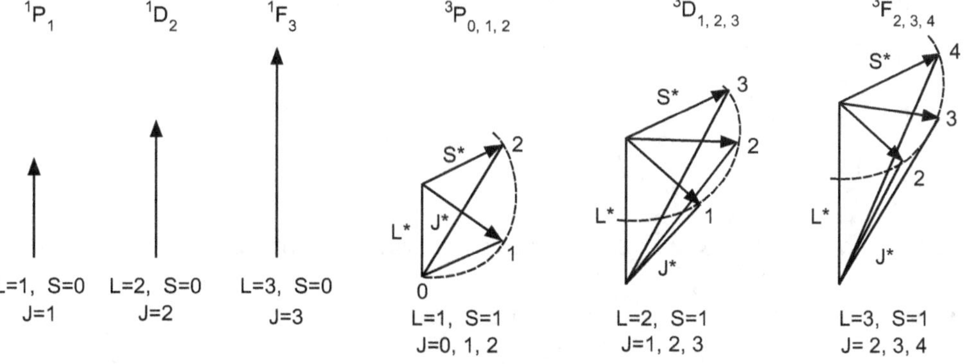

Fig. 3.5 : Vector diagrams of two pd valence electrons in LS coupling

3.1.2 jj Coupling   (Oct. 15)

- In this type of coupling, the spin s_1^* of one electron is quantized with respect to its own l_1^* to form a resultant j_1^* such that j_1 takes half-integral values only.

- Similarly, for second electron, s_2^* and l_2^* give j_2^*, which also takes half-integral values. These j_1^* and j_2^* are in turn quantized with respect to each other to form a resultant J^*, such that J takes integral values only. That is

$$j_1^* = l_1^* + s_1^*$$

and

$$j_2^* = l_2^* + s_2^*$$

The resultant is

$$J^* = j_1^* + j_2^*$$

The jj coupling scheme is represented in Fig. 3.6.

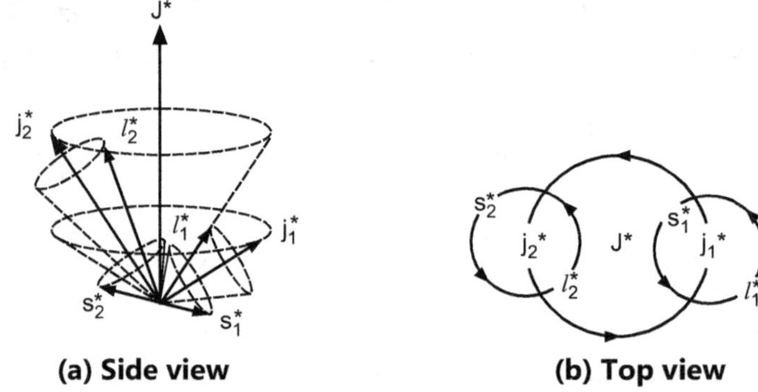

(a) Side view   **(b) Top view**

Fig. 3.6 : jj coupling of two electrons

As an example, consider 'pd' configuration.

For the 'p' electron :

$$l_1 = 1, \quad s_1 = \frac{1}{2} \quad \therefore \ j_1 = \frac{1}{2} \text{ and } \frac{3}{2}$$

For the 'd' electron :

$$l_2 = 2, \quad s_2 = \frac{1}{2} \quad \therefore \ j_2 = \frac{3}{2} \text{ and } \frac{5}{2}$$

Combining these four values of j in all possible ways, we obtain,

when $\quad j_1 = \frac{1}{2}, \quad j_2 = \frac{3}{2} \quad \therefore \ J = 1 \text{ and } 2$

$\qquad j_1 = \frac{1}{2}, \quad j_2 = \frac{5}{2} \quad \therefore \ J = 2 \text{ and } 3$

$\qquad j_1 = \frac{3}{2}, \quad j_2 = \frac{3}{2} \quad \therefore \ J = 0, 1, 2, 3$

$\qquad j_1 = \frac{3}{2}, \quad j_2 = \frac{5}{2} \quad \therefore \ J = 1, 2, 3, 4, \text{ etc.}$

Thus, in all 'J' values are twelve, giving the same number of terms arrived in the LS coupling scheme. The vector diagrams of 'pd' electrons in jj coupling are shown in Fig. 3.7.

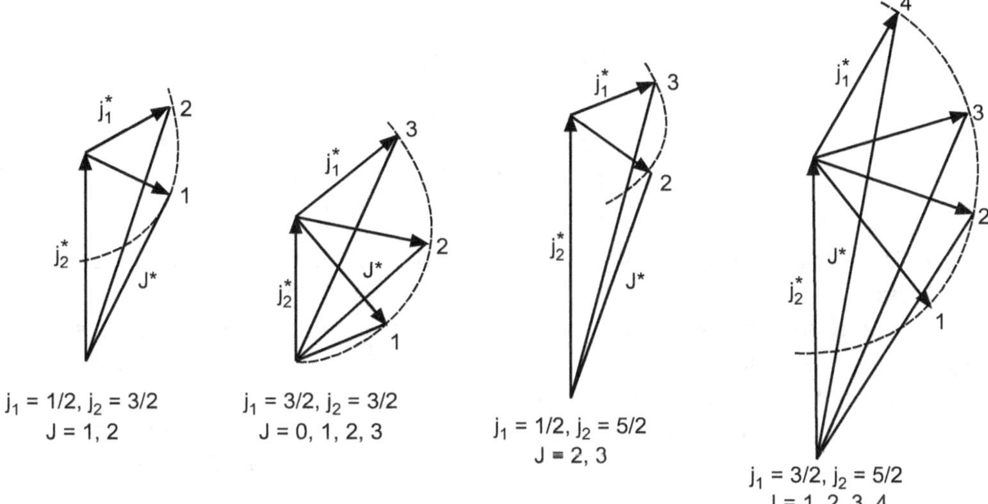

Fig. 3.7: Vector diagrams of two pd valence electrons in jj coupling

3.2 Interaction Energies of LS Couplings   (April 16; Oct. 17, 16)

- We have seen in the case of single-valence electron that due to the interaction between orbital motion and spin motion of the electron, there is a change in the total energy of the system. This energy change gives rise to fine structure of spectral lines.

- Now, in case of two-valence electrons, there are four angular momenta l_1^*, l_2^*, s_1^* and s_2^*, with six possible interactions as -

$$s_1^* \text{ with } s_2^* \qquad l_2^* \text{ with } s_2^*,$$

$$l_1^* \text{ with } l_2^* \qquad l_1^* \text{ with } s_2^*$$

$$l_1^* \text{ with } s_1^* \qquad l_2^* \text{ with } s_1^*$$

- We have seen that the shift of each fine structure level is given by,

$$- \Delta T_{l,\,s} = a\, l^* s^* \cos\, \overline{(l^* s^*)}$$

$$= a\left(\frac{j^{*2} - l^{*2} - s^{*2}}{2}\right) = \Gamma \qquad \qquad \ldots (3.1)$$

Applying this equation to these six interactions, the six energy relations will be

$$\Gamma_1 \;=\; a_1\, s_1^*\, s_2^*\, \cos\,(s_1^*\, s_2^*)$$
$$\Gamma_2 \;=\; a_2\, l_1^*\, l_2^*\, \cos\,(l_1^*\, l_2^*)$$
$$\Gamma_3 \;=\; a_3\, l_1^*\, s_1^*\, \cos\,(l_1^*\, s_1^*)$$
$$\Gamma_4 \;=\; a_4\, l_2^*\, s_2^*\, \cos\,(l_2^*\, s_2^*) \qquad\qquad \dots (3.2)$$
$$\Gamma_5 \;=\; a_5\, l_1^*\, s_2^*\, \cos\,(l_1^*\, s_2^*)$$
$$\Gamma_6 \;=\; a_6\, l_2^*\, s_1^*\, \cos\,(l_2^*\, s_1^*)$$

- According to classical model of the precession of vectors, each spin s^* and each orbit l^* produces a field. The other vectors tend to carry out 'Larmor precession' around this field. The rate of precession is different for different vectors. The vector precessing more rapidly has predominating interaction over other slowly precessing vectors. LS and jj coupling models give good agreement with experiments.

- In LS coupling, the interaction energies Γ_1 and Γ_2 are assumed to predominate over Γ_3 and Γ_4, while Γ_5 and Γ_6 are assumed negligibly small. The vector diagram of this coupling scheme is shown in Fig. 3.8 with Γ_1 and Γ_2 predominant, s_1^* and s_2^* precess rapidly around their resultant S^* and l_1^* and l_2^* precess rapidly around their resultant L^*. Due to weaker interactions Γ_3 and Γ_4, L^* and S^* precess more slowly around their resultant J^*. We consider the terms

$$\Gamma_1 = a_1\, s_1^*\, s_2^*\, \cos\,(s_1^*\, s_2^*)\;; \qquad\qquad \Gamma_3 = a_3\, l_1^*\, s_1^*\, \cos\,(l_1^*\, s_1^*)$$
$$\Gamma_2 = a_2\, l_1^*\, l_2^*\, \cos\,(l_1^*\, l_2^*)\;; \qquad\qquad \Gamma_4 = a_4\, l_2^*\, s_2^*\, \cos\,(l_2^*\, s_2^*)$$

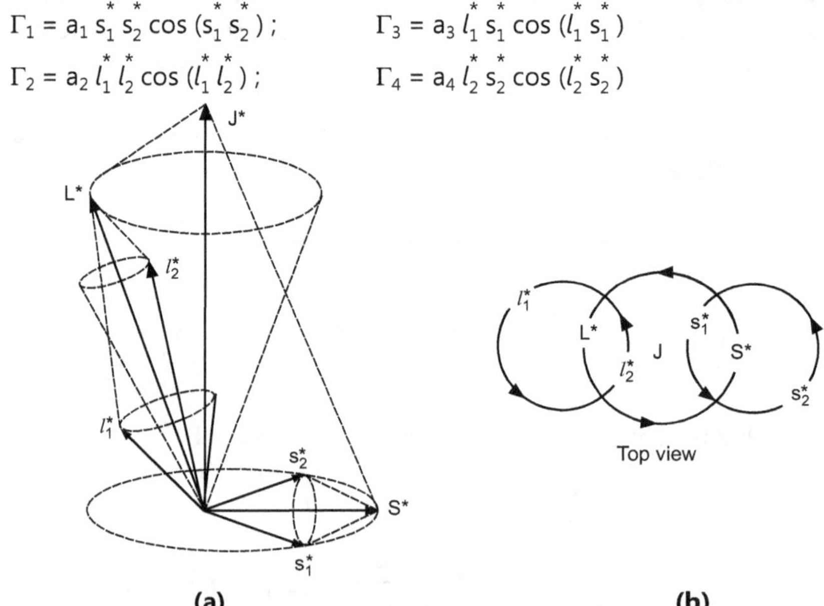

Fig. 3.8 : Sketch of vector model for L - S coupling

By cosine law for triangles,

$$S^{*2} = s_1^{*2} + s_2^{*2} + 2\,s_1^*\,s_2^*\,\cos(s_1^*\,s_2^*) \qquad \text{... (3.3)}$$

$$\therefore \quad s_1^*\,s_2^*\,\cos(s_1^*\,s_2^*) = \frac{S^{*2} - s_1^{*2} - s_2^{*2}}{2} \qquad \text{... (3.4)}$$

Hence,

$$\Gamma_1 = \frac{1}{2}\,a_1\,(S^{*2} - s_1^{*2} - s_2^{*2}) \qquad \text{... (3.5)}$$

and

$$\Gamma_2 = \frac{1}{2}\,a_2\,(L^{*2} - l_1^{*2} - l_2^{*2}) \qquad \text{... (3.6)}$$

- The average value of the cosine must be evaluated as the angles between the vectors are continuously changing. Due to constancy of certain angles throughout the various precessions, the average value of cosine can be easily calculated. A well-known theorem in trigonometry gives the average value in terms of projections.

- For example, the projection of l_1^* on L^*, the L^* on S^* and finally S^* on s_1^* gives, for the average cosine.

- This average value is

$$\overline{\cos(l_1^*\,s_1^*)} = \cos(l_1^*\,L^*)\,\cos(L^*\,S^*)\,\cos(S^*\,s_1^*) \qquad \text{... (3.7)}$$

Similarly,

$$\overline{\cos(l_2^*\,s_2^*)} = \cos(l_2^*\,L^*)\,\cos(L^*\,S^*)\,\cos(S^*\,s_2^*) \qquad \text{... (3.8)}$$

- It is assumed that s_1^*, l_1^* precess much more rapidly than L^* and S^*, so that their components normal to the respective axes of rotation will cancel out.

- Using cosine law for vector triangle of L^*S^*, average values of cosines, taking common factor $\cos(L^*S^*)$, we get

$$\Gamma_3 + \Gamma_4 = (a_3\,\alpha_3 + a_4\,\alpha_4)\,L^*S^*\cos(L^*S^*) \qquad \text{... (3.9)}$$

$$= \frac{1}{2}\,(a_3\,\alpha_3 + a_4\,\alpha_4)\,(J^{*2} - L^{*2} - S^{*2}) \qquad \text{... (3.10)}$$

where

$$\alpha_3 = \frac{s_1^{*2} - s_2^{*2} + S^{*2}}{2\,S^{*2}} \cdot \frac{l_1^{*2} - l_2^{*2} + L^{*2}}{2\,L^{*2}} \qquad \text{... (3.11)}$$

and

$$\alpha_4 = \frac{s_2^{*2} - s_1^{*2} + S^{*2}}{2\,S^{*2}} \cdot \frac{l_2^{*2} - l_1^{*2} + L^{*2}}{2\,L^{*2}} \qquad \text{... (3.12)}$$

- For any given triplet, s_1^*, s_2^*, l_1^*, l_2^*, S^* and L^* are fixed in magnitude, hence a_3, a_4, α_3 and α_4 are constants. Hence, writing

$$A = a_3\,\alpha_3 + a_4\,\alpha_4 + \ldots \qquad \text{... (3.13)}$$

Equation (3.10) becomes

$$\Gamma_3 + \Gamma_4 = A\,L^*S^*\cos(L^*S^*) = \frac{1}{2}\,A\,(J^{*2} - L^{*2} - S^{*2}) \qquad \text{... (3.14)}$$

- It is necessary to evaluate the 'A' coefficients to obtain complete expression of energies.

- From the comparison of calculated energies and observed fine structure, it is found that spin-orbit interactions are due to electrostatic effects. It is the general observation that the singlet levels lie above the corresponding triplet levels of the same electron configuration. Hence, the coefficient 'a_1' of Γ_1 is negative. Now, we are in position to calculate the effect of each interaction Γ_1, Γ_2, Γ_3, Γ_4.

pd configuration :

- Let us consider the 'pd' electron configuration. In this configuration the first electron is in p subshell and second electron is in d subshell. Therefore, $l_1 = 1$, $s_1 = \dfrac{1}{2}$ and $l_2 = 2$, $s_2 = \dfrac{1}{2}$.

The equation for Γ_1 is

$$\Gamma_1 = \frac{1}{2} a_1 (S^{*2} - s_1^{*2} - s_2^{*2}) \qquad \ldots (3.15)$$

Here,

$$s_1 = \frac{1}{2},\ s_2 = \frac{1}{2} \ \therefore \ S = 0 \text{ or } 1.$$

S = 0 gives singlets and S = 1 gives triplets.

$\therefore$ For S = 0,

$$\Gamma_1 = \frac{1}{2} a_1 \left[0 - \left(\sqrt{\frac{3}{4}}\right)^2 - \left(\sqrt{\frac{3}{4}}\right)^2 \right]$$

$\therefore$

$$\Gamma_1 = \frac{1}{2} a_1 \left(-\frac{3}{2}\right) = -\frac{3a_1}{4}$$

For S = 1,

$$\Gamma_1 = \frac{1}{2} a_1 \left[(\sqrt{2})^2 - \left(\sqrt{\frac{3}{4}}\right)^2 - \left(\sqrt{\frac{3}{4}}\right)^2 \right]$$

$$\Gamma_1 = \frac{1}{2} a_1 \left(2 - \frac{3}{2}\right) = \frac{a_1}{4}$$

Thus,

$$\Gamma_1 = -\frac{3a_1}{4} \text{ and } \frac{a_1}{4}$$

This is the first splitting shown in Fig. 3.9. The equation for Γ_2 is

$$\Gamma_2 = \frac{1}{2} a_2 (L^{*2} - l_1^{*2} - l_2^{*2})$$

Here $l_1 = 1$, $l_2 = 2$

$\therefore$ L = 1, 2, 3 i.e. P, D, F terms respectively.

For L = 1,　　　　$\Gamma_2 = \dfrac{1}{2} a_2 [(\sqrt{2})^2 - (\sqrt{2})^2 - (\sqrt{6})^2]$

$\therefore$　　　　$\Gamma_2 = \dfrac{1}{2} a_2 (-6) = -3 a_2$

For L = 2,　　　　$\Gamma_2 = \dfrac{1}{2} a_2 [(\sqrt{6})^2 - (\sqrt{2})^2 - (\sqrt{6})^2] = \dfrac{1}{2} a_2 (-2)$

$\therefore$　　　　$\Gamma_2 = -a_2$

For L = 3,　　　　$\Gamma_2 = \dfrac{1}{2} a_2 [(\sqrt{12})^2 - (\sqrt{2})^2 - (\sqrt{6})^2]$

$\therefore$　　　　$\Gamma_2 = \dfrac{1}{2} a_2 (12 - 2 - 6) = 2 a_2$

- Thus, the values of Γ_2 are $-3a_2$, $-a_2$ and $2a_2$. This is the second splitting shown in Fig. 3.9 for singlets and triplets.

- The Hund's rule applied for LS coupling states that out of all terms with the same L value arising from any given electron configuration,

  (i)　a term with the highest multiplicity i.e. the highest S value, will lie deepest.

  (ii)　of these the terms with the highest L value will lie deepest.

- The separation of the terms from the centre of gravity terms can be calculated from $\Gamma_3 + \Gamma_4$.

$\therefore$　　　　$\Gamma_3 + \Gamma_4 = \dfrac{1}{2} A (J^{*2} - L^{*2} - S^{*2})$

or　　　　$\Gamma_3 + \Gamma_4 = \dfrac{1}{2} A [J(J+1) - L(L+1) - S(S+1)]$

For the term $^3P_0$,　　　　$\Gamma_3 + \Gamma_4 = \dfrac{1}{2} A [0(0+1) - 1(1+1) - 1(1+1)] = -2 A$

For the term $^3P_1$　　　　$\Gamma_3 + \Gamma_4 = \dfrac{1}{2} A [1(1+1) - 1(1+1) - 1(1+1)] = -A$

For the term $^3P_2$　　　　$\Gamma_3 + \Gamma_4 = \dfrac{1}{2} A [2(2+1) - 1(1+1) - 1(1+1)] = A$

Similarly, we can find $\Gamma_3 + \Gamma_4$ for $^3D_{1,2,3}$ and $^3F_{2,3,4}$ terms.

For the singlet terms with S = 0, the α's and hence A's vanish.

- In similar way, all the separations shown in the diagram (Fig. 3.9) can be calculated. The schematic representation of the fine structure of pd configuration is shown in Fig. 3.9.

- In this figure, energy level attributed to n and l values of the two electrons is shown at the left.

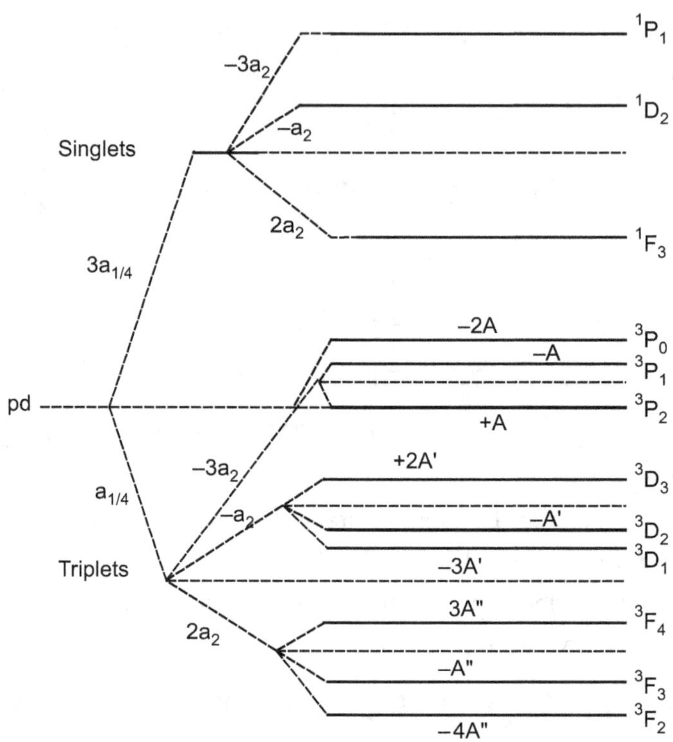

Fig. 3.9 : Fine structure separations of pd configuration in LS coupling

sp configuration :

In 'sp' configuration the first electron is in s subshell and second in p subshell. Therefore, $l_1 = 0$, $s_1 = \dfrac{1}{2}$ and $l_2 = 12$, $s_2 = \dfrac{1}{2}$.

The expression for tau factor, is,

$$\Gamma_1 = \frac{1}{2} a_1 \left(S^{*2} - s_1^{*2} - s_2^{*2} \right)$$

For S = 0, $$\Gamma_1 = -\frac{3a_1}{4}$$

For S = 1, $$\Gamma_1 = \frac{a_1}{4}$$

Thus, the values of Γ_1 are $-\dfrac{3a_1}{4}$ and $\dfrac{a_1}{4}$. This is the first splitting shown in Fig. 3.10.

Now, as $l_1 = 0$, $l_2 = 1$, we have L = 1, i.e. only 'p' term.

For L = 1, $$\Gamma_2 = \frac{1}{2} a_2 [(\sqrt{2})^2 - 0 - (\sqrt{2})^2] = \frac{1}{2} a_2 [2 - 0 - 2] = 0$$

$\therefore$ $$\Gamma_2 = 0$$

This is the second splitting shown in Fig. 3.10 for singlets and triplets.

For $S = 1$ and $L = 1$, there are three triplet terms $^3P_0$, $^3P_1$, and $^3P_2$. The separation between them can be obtained by calculating $\Gamma_3 + \Gamma_4$.

For $^3P_0$ term,

$$\Gamma_3 + \Gamma_4 = \frac{1}{2} A [0(0+1) - 1(1+1) - 1(1+1)] = -2A$$

Similarly,

For $^3P_1$ term, $\Gamma_3 + \Gamma_4 = -A$ and for $^3P_2$ term, $\Gamma_3 + \Gamma_4 = +A$.

All the separations shown in Fig. 3.10 can be calculated in this way.

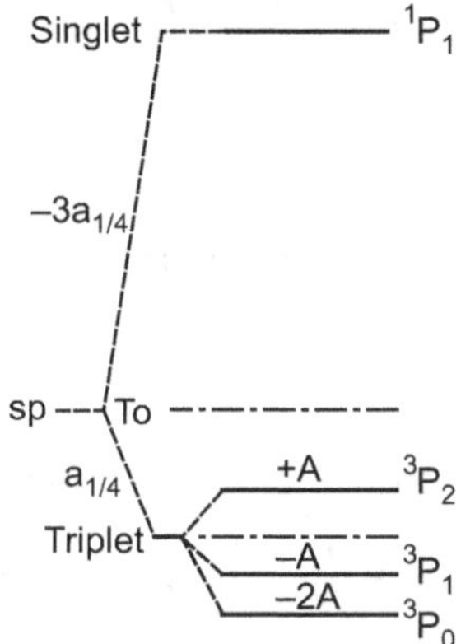

Fig. 3.10 : Schematic representation of the interaction energies between two valence electrons in LS coupling

3.3 The Lande Interval Rule   (Oct. 15, April 17)

Statement : For a given triplet i.e. for given 'S' and 'L', each fine structure term difference is proportional to the larger of the two J values associated with it. i.e. the separation in the energy of adjacent levels of a multiplet is proportional to the total angular momentum quantum number of the level of higher energy. This prediction is called the Lande interval rule.

- The rule is of great help in determining the values of J that are to be assigned to various observed levels. This rule is widely used in atomic physics.

- In LS coupling, we have already seen that $\Gamma_3 + \Gamma_4$ give the separation among the components of triplets and it is given by

$$\Gamma_3 + \Gamma_4 = \frac{1}{2} A [J (J + 1) - L (L + 1) - S (S + 1)] \qquad \text{... (3.16)}$$

- Let the adjacent terms in a given triplet have J values as J and $(J + 1)$.

$\therefore$ Term difference

$$- \Delta T = \frac{1}{2} A [(J + 1)(J + 2) - L(L + 1) - S(S + 1)]$$

$$- \frac{1}{2} A [J(J + 1) - L(L + 1) - S(S + 1)]$$

$\therefore$
$$\Delta T = \frac{1}{2} A [(J + 1)(J + 2) - J(J + 1)]$$

$\therefore$
$$\Delta T = A(J + 1) \qquad \qquad \text{... (3.17)}$$

- The intervals are thus proportional to the larger J values. This is an excellent agreement between observed separations and the interval rule. This clearly justifies the use of the vector atom model and signifies the LS electron coupling. The rule also determines the separation of fine structure lines in LS coupling.
- The same result can be obtained by considering pd configuration. Now, consider pd configuration.

  Three triplets are : $^3P_0\ ^3P_1\ ^3P_2$, $^3D_1\ ^3D_2\ ^3D_3$, $^3F_2\ ^3F_3\ ^3F_4$

  In this the ratios are : $1:2,\ 2:3,\ 3:4$
- The numerical ratios conclude that term difference is proportional to larger of the two J values.

  A graphical representation of the Lande interval rule is shown in Fig. 3.11 for ^3D term.

  For ^3D term : $S = 1, L = 2, J = 1, 2, 3$.

  Thus, the components of given triplet ^3D are $^3D_1\ ^3D_2$ and $^3D_3$.

Now for $^3D_1$,
$$\Gamma_3 + \Gamma_4 = \frac{A}{2} [J(J + 1) - L(L + 1) - S(S + 1)]$$

$$= \frac{A}{2} [1(1 + 1) - 2(2 + 1) - 1(1 + 1)]$$

$\therefore$
$$\Gamma_3 + \Gamma_4 = -3A \qquad \qquad \text{... (3.18)}$$

Similarly, for $^3D_2$,
$$\Gamma_3 + \Gamma_4 = \frac{A}{2} [6 - 6 - 2] = -A \qquad \qquad \text{... (3.19)}$$

and for $^3D_3$ term,
$$\Gamma_3 + \Gamma_4 = \frac{A}{2} [12 - 6 - 2] = 2A \qquad \qquad \text{... (3.20)}$$

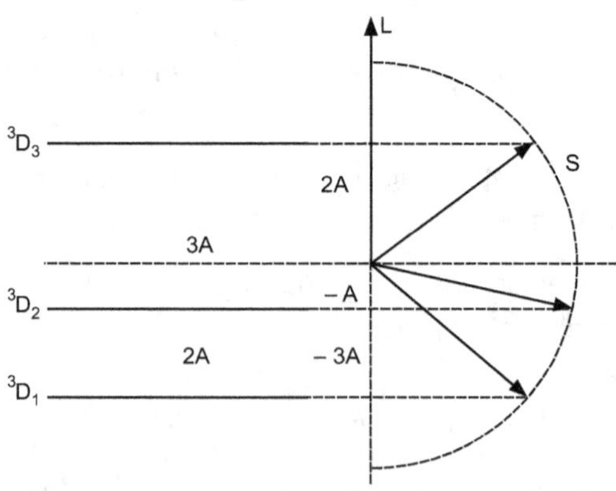

Fig. 3.11 : Graphical representation of Lande interval rule for ^3D term

- Thus, in Fig. 3.11, S^* is shown quantized with L^* in three allowed positions J = 1, 2 and 3. The resultant intervals have the ratio 2 : 3. The dotted line is at the centre of gravity. The J vectors are not shown in the representation.

3.4 Spectra of Helium

- Helium (Z = 2) has two electrons, both of which according to Pauli's principle, can occupy the first K shell. The two electrons complete the first period of the periodic system. The normal helium atom is therefore represented by $1s^2$ i.e. two electrons, for which n = 1 and l = 0. The total number being affixed at the top of s. A single electron is responsible for the energy levels of both hydrogen and sodium. However, there are two 1s electrons in the ground state of helium and coupling affects the properties and behaviour of the helium atom.

- Helium atom contains two valence electrons. Therefore its emission spectrum is of the same type as are given by alkaline earths. The spectrogram for helium shows two systems- one singlet system containing all sharp, principal, diffuse and fundamental series, and one triplet system which also contains all these four chief series. The singlet system belongs to para-helium (spins antiparallel) because for singlets S = 0. The triplet system belongs to ortho-helium (spins parallel) because S = 1. The singlet terms arise when one electron remains in 1s level and second electron is excited to higher levels with spin antiparallel to first electron.

- The various levels represent configurations in which one electron is in its ground state and the other is in an excited state. There is a division into singlet and triplet states (i.e. para-helium and ortho-helium) as shown in Fig. 3.12. An ortho-helium atom can loose excitation energy in a collision and become one of para-helium, while a para-helium atom can gain excitation energy in a collision and become one of ortho-helium. Ordinary liquid or gaseous helium is therefore a mixture of both. The lowest singlet state is 1s1s ($^1S_0$). The lowest triplet states are metastable.

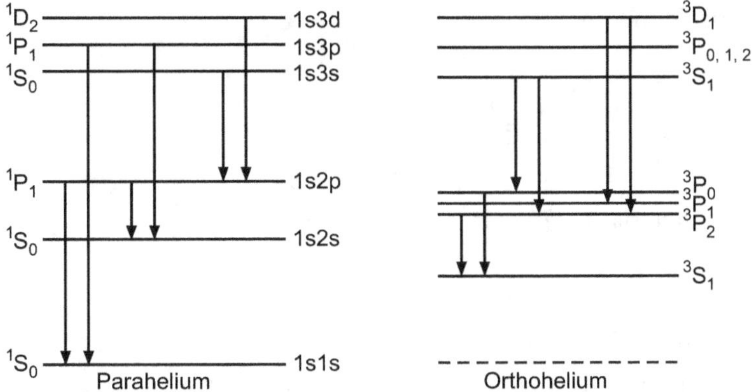

Fig. 3.12 : Terms and transitions in the spectrum of He

- Ortho-helium does not exist in the state defined by the principal quantum number $n = 1$ (1^3S_1). This is due to Pauli's principle since the two electrons would have parallel spins and therefore identical sets of quantum numbers. The lowest state of ortho-helium is 2^3S_1. The state 3^2S_1 represents metastable state. The 2^1S_0 state is also metastable state, because of selection rule $\Delta l = \pm 1$ restricts any transition from this state to 1^1S_o state.

- The diagram shows that there is enormous energy shift from the normal $1s^2\ {}^1S_0$ state to the first excited states $1s\ 2s\ {}^1S_0$ and ${}^3S_1$. It is the characteristic of helium and all inert gases. The fine structure of the very narrow ortho-helium series triplets was predicted by Slater and verified by Heisenberg. Heisenberg has shown that the enormous energy difference between the first two or three triplet levels and their associated singlet levels is due to an electrostatic resonance interaction. The normal state of helium, for example to be 24.470 volts below the series limit, a value differing only 0.003 volt from the spectroscopic value 24.467 volts.

Solved Problems

Problem 3.1 : *Using 'p-p' electron configuration and 'd-d' electron configuration, find all possible values of the total angular momenta.*

Solution : (1) For p-p electron configuration :

For first electron :

$$l_1 = 1, s_1 = \frac{1}{2}$$

For second electron :

$$l_2 = 1,\ s_2 = \frac{1}{2}$$

∴ Resultant orbital momentum is,

$$L = (l_1 + l_2) \text{ to } (l_1 - l_2) = 0, 1, 2$$

and resultant spin momentum is

$$S = (s_1 + s_2) \text{ to } (s_1 - s_2)$$
$$S = 0, 1$$

∴ Total angular momentum is

$$J = L + S \text{ to } L - S$$

∴ For $S = 0$, multiplicity $(2S + 1) = 1$

∴
$$J = L = 0, 1, 2$$

∴ The singlet terms are ${}^1S_0,\ {}^1P_1,\ {}^1D_2$

∴ For $S = 1$, multiplicity $(2S + 1) = 3$

For $L = 0$, $J = 1$; for $L = 1$, $J = 0, 1, 2$ and for $L = 2$, $J = 1, 2, 3$

∴ The singlet and triplet terms are ${}^3S_1,\ {}^3P_{0,\,1,\,2},\ {}^3D_{1,\,2,\,3}$

(2) For d-d electron configuration :

For first electron :

$$l_1 = 2, \quad s_1 = \frac{1}{2}$$

For second electron :

$$l_2 = 2, \quad s_2 = \frac{1}{2}$$

∴   Resultant orbital momentum is,

$$L = (l_1 + l_2) \text{ to } (l_1 - l_2) = 0, 1, 2, 3, 4$$

and resultant spin momentum is

$$S = (s_1 + s_2) \text{ to } (s_1 - s_2)$$

∴ $$S = 0, 1$$

∴   Total angular momentum is

$$J = L + S \text{ to } L - S$$

For $S = 0$, $m = 2S + 1 = 1$

∴ $$J = L = 0, 1, 2, 3, 4$$

∴   The singlet terms are $^1S_0$, $^1P_1$, $^1D_2$, $^1F_3$, $^1G_4$

For $S = 1$, $m = 2S + 1 = 3$

When $L = 0$, $J = 1$

When $L = 1$, $J = 0, 1, 2$

       $L = 2$, $J = 1, 2, 3$

       $L = 3$, $J = 2, 3, 4$

       $L = 4$, $J = 3, 4, 5$

∴   The singlet and triplet terms are $^3S_1$, $^3P_{0, 1, 2}$, $^3D_{1, 2, 3}$, $^3F_{2, 3, 4}$, $^3G_{3, 4, 5}$.   ... **Ans.**

Problem 3.2 : *Determine the values of L, the quantum number describing the ll coupling of the angular momenta of two atomic orbitals, for two d electrons. What are the corresponding letter symbols ?*

Solution : The permitted values of L are given by

$$L = (l_1 + l_2), l_1 + l_2 - 1, \dots |l_1 - l_2|$$

For two d electrons, $l_1 = 2$, $l_2 = 2$

∴ $$L = 2 + 2 = 4$$
$$L = 2 + 2 - 1 = 3$$
$$L = 2 + 2 - 2 = 2$$
$$L = 2 + 2 - 3 = 1$$
$$L = 2 + 2 - 4 = |2 - 2| = 0 \qquad \text{... \textbf{Ans.}}$$

which correspond to the letters G, F, D, P and S respectively.

Problem 3.3 : *In an atom, the components of a normal triplet have separation of 19 and 38 cm⁻¹ between adjacent levels in LS coupling. The next higher state in multiplet has separations 22 and 33 cm⁻¹ respectively. Determine the terms for these two states and draw the corresponding energy level diagram showing the allowed transitions.*

Solution : According to Lande interval rule, the separation between adjacent levels is proportional to the higher J value. Let first multiplet have J values J, J + 1 and J + 2.

$$\therefore \qquad\qquad A\,(J + 1) = 19$$

and $\qquad\qquad A\,(J + 2) = 38$ where 'A' is constant.

This gives J = 0

∴ The states have J values 0, 1, 2.

Let the second multiplet have J values J', J' + 1 and J' + 2.

$$\therefore \qquad\qquad A'\,(J' + 1) = 22$$

$$A'\,(J' + 2) = 33$$

where A' is constant.

This gives J' = 1.

∴ The states have J values 1, 2, 3.

Now, S = 1 ∴ L = 2

∴ Terms are $^3D_3$, $^3D_2$ and $^3D_1$.

∴ According to selection rule,

$$\Delta J = 0, \pm 1 \;(0 \rightarrow 0 \text{ excluded})$$ **... Ans.**

∴ The possible allowed transitions are 6 and shown in Fig. 3.13.

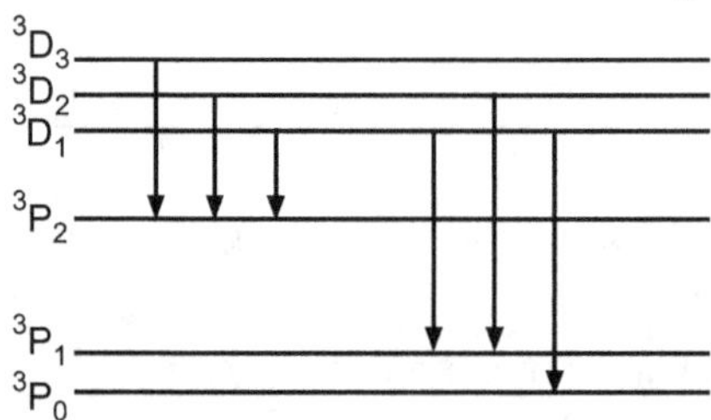

Fig. 3.13 : Triplet-triplet transitions

Summary

1. There are several ways in which different vectors of electrons may combine to give the vectors representing the atom as a whole. They are *ll* coupling, ss coupling, jj coupling and LS coupling.

2. If two electrons have the same quantum number and same azimuthal quantum number, they are called equivalent electrons.

3. In case of two-valence electrons, there are four angular momenta $\vec{l_1}$, $\vec{l_2}$, $\vec{s_1}$ and $\vec{s_2}$, with six possible interactions as - $\vec{s_1}$ with $\vec{s_2}$, $\vec{l_2}$ with $\vec{s_2}$, $\vec{l_1}$ with $\vec{l_2}$, $\vec{l_1}$ with $\vec{s_2}$

$\vec{l_1}$ with $\vec{s_1}$ and $\vec{l_2}$ with $\vec{s_1}$

3. For a given triplet for given S and L each fine structure term difference is proportional to the larger of the two J values associated with it i.e. the separation of energy of adjacent levels of multiplet is proportional to the total angular momentum quantum number of the level of higher energy. This is called Lande interval rule.

Exercise

(A) Short Answer Type Questions :

1. Draw vector diagrams of LS and jj couplings.

2   What are tau (Γ) factors ?

3. What are the odd terms and even terms ?

4. Define equivalent electrons.

5. Determine the different values for the total orbital quantum number of two-electron systems with $L_1 = 3$ and $L_2 = 2$.

6. What are the 'L' and 'S' quantum numbers corresponding to $^3D_2$?

7. Determine the values of J, the quantum number describing the Russel-Saunder's coupling between 'L' and 'S' for two 'd' electrons.

(B) Long Answer Type Questions :

1. Explain LS coupling scheme for two-valence electron systems using neat vector diagram.

2. Explain jj coupling scheme for two-valence electron systems using neat vector diagram.

3. State and explain Lande interval rule. Represent it graphically for ^3D term.

4. Discuss spectra of Helium. Mention the characteristics of this spectra.

5. Give an account of the spectral terms arising from two equivalent electrons.

6. Describe the broad features of the spectrum of Helium atom and show how the observations have been accounted for.

7. Write short notes on the following :

    (i)   Coupling schemes,

    (ii)   Singlet and triplet series in two-valence electron atom,

    (iii) Lande interval rule,

    (iv)   Spectra of He atom.

(C) Numerical Problems :

1. Determine singlet-triplet separations in terms of interaction energies between two valence electrons in 'sp' configuration. (Use LS coupling).

2. Determine singlet-triplet separations in terms of interaction energies between two valence electron in 'pd' configuration. (Use LS coupling).

3. Derive all the terms arising from the electron configuration f-g using L-S coupling scheme.

4. Using vector diagrams, determine the possible values of the total angular momentum of a two-electron system for which $l_1 = 3$ and $l_2 = 2$.

5. Using vector diagrams, determine the possible values of the total angular momentum of an electron system for which (a) L = 2, S = 1, (b) L = 3, S = 1.

6. Determine possible values of $\Gamma_3 + \Gamma_4$ for L = 2 and S = 1.

7. Give the graphical representation of Lande interval rule for ^3P term.

Chapter **4**...

Zeeman Effect

Contents ...

Pieter Zeeman

Pieter Zeeman (Dutch: 25 May 1865 – 9 October 1943) was a Dutch physicist who shared the 1902 Nobel Prize in Physics with Hendrik Lorentz for his discovery of the Zeeman effect. Zeeman effect is splitting of atomic energy levels and associated spectral lines when the atoms are placed in magnetic field. Zeeman effect is very important in NMR, ESR spectroscopy, MRI and Mössbauer spectroscopy.

Introduction

- The effect of magnetic field on the spectrum was first discovered by Zeeman in 1896 and is called Zeeman effect after his name. Primarily it was observed that a single spectral line splits up into three components, one line has got a large frequency, other a lower frequency than frequency of original line. The third line has the frequency of original line. In 1913, Stark studied the effect of electric field on spectral line and was called as Stark effect.

- In present topic we shall discuss these effects.

4.1 Early Discoveries and Developments

- In 1896, Zeeman discovered that when a sodium flame is kept between the poles of powerful electromagnet, two lines of first principal doublet are significantly broadened. Satisfactory explanation of this Zeeman effect was given by Lorentz on the basis of electron theory. He predicted that light from these lines should be polarized by the magnetic field. It should be plane polarized if viewed at right angles to the field and circularly polarized if viewed in the direction parallel to lines of force. These predictions were verified by Zeeman by means of Nicol prism as analyzer later on.

- According to Lorentz's theory when a light source is placed in magnetic field, the motion of the electron is modified in such a way that their period of motion gets charged. In case of electron in circular orbit, plane which is normal to the magnetic field direction H, the electrons will be either speeded up or slowed down by certain amount which depends on H, mass (m), charge on electron (e) and velocity of light (c).

- If υ_o is the orbital frequency of electron without field, the frequency in presence of field is given by $\upsilon_o \pm \Delta\upsilon$, where

$$\Delta\upsilon = \frac{eH}{4\pi mc}$$

- If field is normal to and up from this page, then electrons moving in anticlockwise direction in the plane of paper are speeded by $\Delta\upsilon$. Those electrons moving in a clockwise direction are slowed down by amount $\Delta\upsilon$.

- Consider motion of an electron in an orbit. Resolve the motion into three components along three mutually perpendicular axes. [Refer Fig. 4.1 (a)]. The motion of electron here is pictured as consisting of three harmonic motions each along X, Y and Z axes. If now, in absence of field, electrons are emitting light and we observe radiation along X-direction, only light in y and z motion will be observed. In absence of field, light observed in any direction is unpolarized.

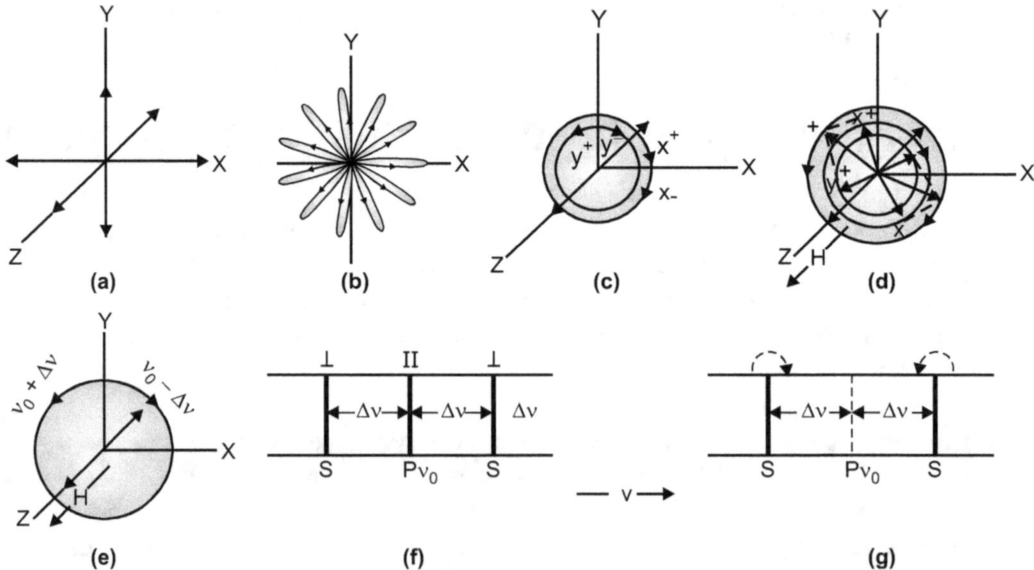

Fig. 4.1 : Schematic diagrams for the classical explanation of the normal Zeeman effect

- If we apply magnetic field along Z-axis, the X and Y motion will be modified and Z motion remains unchanged. Moving perpendicular to field, the X and Y motion will take the form of rosettes as shown in Fig. 4.1 (b). These apparently complex motions can be described in terms of circular motion as follows.

- The SHM given by y-component of Fig. 4.1 (a) for example is equivalent to equal but opposite circular motions y^+ and y^- as shown in Fig. 4.1 (c). Similarly x-component will be represented by two opposite circular motions x^+ and x^-.

- When field is applied x^+ and y^+ rotations will be speeded up by amount $\Delta\upsilon$ and x^- and y^- rotations will be slowed down by $\Delta\upsilon$. Faster x^+ motion is combined with slower x^- motion giving rise to motion as shown in Fig. 4.1 (c).

- The x^+ motion and y^+ motion and x^- and y^- motions are now combined as shown in Fig. 4.1 (d) to form plus and minus resultant. Thus motion of an electron in magnetic field is represented by linear motion along field direction with frequency υ_0 and two circular motions at right angles to this with frequency $\upsilon_0 + \Delta\upsilon$ and other $\upsilon_0 - \Delta\upsilon$ as shown in Fig. 4.1 (e).

- When these motions are viewed in direction of field only circular motions are observed and these are right and left handed circularly polarized light. [Refer Fig. 4.1 (g)] Since light is a transverse wave motion, the motion along Z-axis will not emit light in the field direction. When viewed perpendicular to field, the z motions are observed as plane polarized light with electric field vibrating parallel to the field. Also circular motion is observed as plane polarized light with electric field vector perpendicular to field. So there are three plane polarized components as shown in Fig. 4.1 (f). The central line is unshifted while two lines are equally spaced from central. This is called normal triplet.

- In early stages, Zeeman was unable to split any lines into doublet or triplet but he found that they are widened and their outside edges were polarized. Later, he was able to photograph the two outer components of life in elements like Zn, Cu, Cd and Sn.

- Preston using greater resolving power and dispersion was able to show that certain lines were splitted into triplets when viewed perpendicular to field but others were split up into as many as four or even six components. He also pointed out that the pattern of all lines (called Zeeman pattern) belonging to the same series of spectrum lines was the same and was characteristic of the series. This is known as Preston law.

- From Lorentz classical treatment, Zeeman shift $\Delta\upsilon$ is given by

$$\Delta\upsilon \;=\; \frac{eH}{4\pi mc^2} = 4.67 \times 10^{-5}\,H\;\text{cm}^{-1} = L\;\text{cm}^{-1}$$

where $\Delta\upsilon$ is wave number,
      H is field in gauss,
      c is velocity of light,
      e is the charge on electron in electrostatic unit

Zeeman pattern showing just three lines with exactly these separations are called normal triplets. (Refer Fig. 4.2)

- All other groups, the complex pattern observed in the chromium spectrum are said to exhibit the anomalous Zeeman effect. Anomalous Zeeman effect study was carried out by Paschen and Runge. Each member of principal series of sodium, copper and silver was observed to have 10 components.

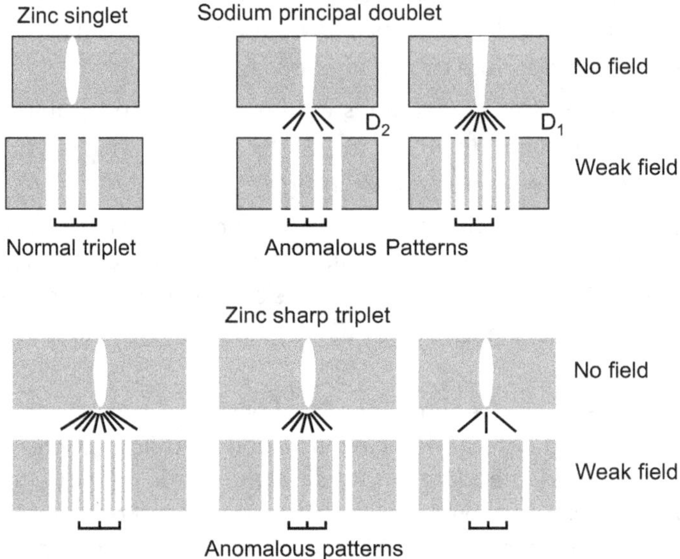

Fig. 4.2 : Normal and anomalous Zeeman effect

(viewed perpendicular to the magnetic field)

- At higher magnetic fields, Zeeman effect ceases to be linear. At even higher temperature, when field strength is comparable to the strength of atom's internal field, electron coupling is disturbed and the spectral lines rearrange. This is called Paschen-Back effect.

4.2 Experimental Arrangement

- Zeeman effect is a magneto-optical phenomenon in which spectral lines are affected by an applied magnetic field and split up into several components.

- The simplest splitting known as normal Zeeman effect is obtained with strong magnetic field. Normal Zeeman effect consists of a triplet when viewed in transverse direction (perpendicular to the direction of magnetic field) and a doublet in longitudinal direction (i.e. parallel to the magnetic field). The more complex resolution of a greater number of lines called anomalous Zeeman effect is produced in weak magnetic field.

- The normal effect can be readily produced and observed with following experimental arrangement.

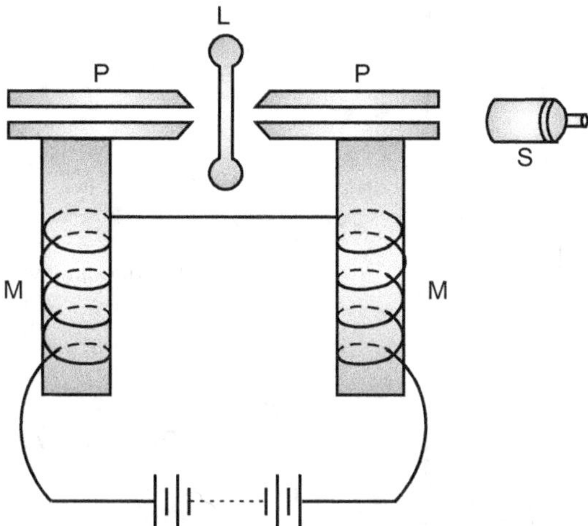

Fig. 4.3 : Experimental set up of normal Zeeman effect

- Zeeman effect experimental set up consists of

  (i)   An electromagnet MM capable of producing a strong magnetic field having a conical pole pieces PP, through which hole have been drilled lengthwise.

  (ii)   Source L (say sodium vapour lamp), emitting spectral lines is kept between pole pieces.

  (iii) High resolving power instrument such as a Lummer Gerke plate with constant deviation spectrometer S is used to observe spectral lines. To obtain a photograph, a camera is used in place of eyepiece of spectrometer.

- When spectral lines are viewed longitudinally through the holes drilled to pole pieces and hence parallel to the direction of field, the lines are found to be split up into two (doublet) one having smaller wavelength than the original. Other having greater wavelength than the original line which is no longer seen. As soon as the current feeding the electromagnet is switched off, two lines disappear and original line alone is seen. Two lines are observed when current is on. These lines are found to be symmetrically situated about the position of parent line so that Zeeman shift (change in wavelength $d\lambda$) is same in both the cases. It has been observed that both these two lines are found to be circularly polarized in opposite directions as shown by arrows above the line.

- When viewed transversely (i.e. perpendicular to the direction of magnetic field) the line becomes triplet, the central one having the same wavelength as original line and the outer one occupying same position as doublets in previous case. All three lines are plane polarized, but vibrations of central line are parallel to magnetic field while those of the outer ones perpendicular to the field.

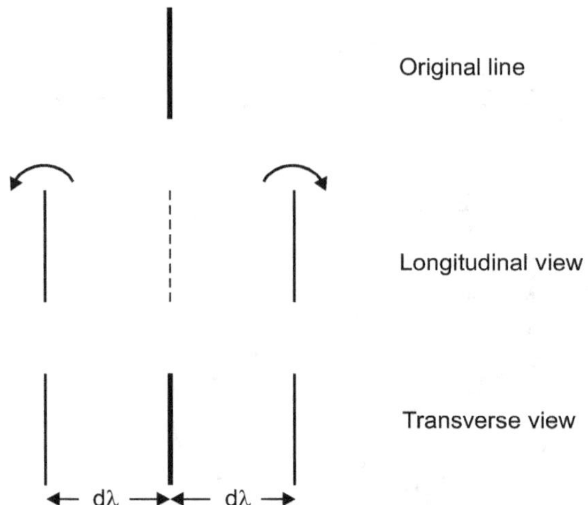

Fig. 4.4 : Zeeman pattern in transverse and longitudinal view

- The resolved line in both the cases can be photographed using a camera in place of eyepiece of spectrometer and from photos obtained Zeeman shift dλ can be measured.

4.3 Normal and Anomalous Zeeman Effect

Normal Zeeman Effect :   **(April 17; Oct. 17, 15)**

- The splitting of spectral lines into three components in a magnetic field when viewed in a direction perpendicular to that of magnetic field is called normal Zeeman effect. Normal Zeeman effect is obtained from sources of elements like Ca, Cu, Zn, Cd, etc.

Anomalous Zeeman Effect :   **(Oct. 17)**

- When a single spectral line is splitted into four or more lines when observed in a direction perpendicular that of magnetic field, it is known as anomalous Zeeman effect.

- It is called anomalous Zeeman effect because it cannot be explained on the basis of classical physics. Anomalous Zeeman effect is obtained from elements like Na, Cr, etc.

Classical Theory of Normal Zeeman Effect :   **(Oct. 15)**

- The normal Zeeman effect has been well explained by Lorentz on the consideration of electronic theory of matter.

- Consider electron revolving in a circular orbit. The centripetal force acting on it is given by

$$F = \frac{mv^2}{r} \qquad \qquad \dots (4.1)$$

where v is velocity of electron in its orbit of radius r.

But, $$v = r\omega = \frac{2\pi}{T} r = 2\pi r\upsilon$$

∴ $$F = \frac{m}{r} \cdot 4\pi^2 r^2 \upsilon^2 = 4\pi^2 mr\upsilon^2 \qquad \qquad \dots (4.2)$$

- If magnetic field is applied in direction perpendicular to the plane of electronic orbit, the additional force = Hev = He$2\pi r\upsilon$ acts on electron.

- According to Fleming's left hand rule, the direction of this force will be towards or away from the centre, according to as electronic motion is clockwise or anticlockwise. If additional force is towards centre of the atom, the orbit contract and frequency of rotation increases.

- If additional force is away from the centre, orbit expands and frequency decreases, then net force is

$$F \pm \text{He}2\pi r\upsilon \; = \; 4\pi^2 \, mr\upsilon_l^2 \qquad\qquad \text{... (4.3)}$$

where $\qquad\qquad \upsilon_l$ = Frequency of orbit

Substituting in equation (4.2) from equation (4.3), we get

$$4\pi^2 \, mr \left(\upsilon_l^2 - \upsilon^2 \right) \; = \; \pm \, \text{He}2\pi r\upsilon$$

$$\upsilon_l^2 - \upsilon^2 \; = \; \pm \frac{\text{He}\upsilon}{2\pi m}$$

$$\left(\upsilon_l + \upsilon \right) \left(\upsilon_l - \upsilon \right) \; = \; \pm \frac{\text{He}\upsilon}{2\pi m}$$

But $\upsilon_l \approx \upsilon$, $\upsilon_l + \upsilon = 2\upsilon$ and $\upsilon_l - \upsilon = d\upsilon$

$$d\upsilon \cdot 2\upsilon \; = \; \pm \frac{\text{He}\upsilon}{2\pi m}$$

$$d\upsilon \; = \; \pm \frac{\text{He}}{4\pi m}$$

$$\upsilon \; = \; \frac{c}{\lambda}$$

$$d\upsilon \; = \; -\frac{c}{\lambda^2} \, d\lambda$$

$$d\lambda \; = \; \pm \frac{\lambda^2}{c} \frac{\text{He}}{4\pi m}$$

where λ is wavelength of original line and $d\lambda$ is change in wavelength or Zeeman effect.

Explanation of normal Zeeman effect on the basis of quantum theory :   **(Oct. 17, 16)**

Atom in a magnetic field :

- When atom is kept in a magnetic field H, a magnetic dipole moment μ exists. This magnetic dipole moment has potential energy δE, that depends upon both the magnitude μ of its magnetic moment and orientation of this moment with respect to field as shown in Fig. 4.5.

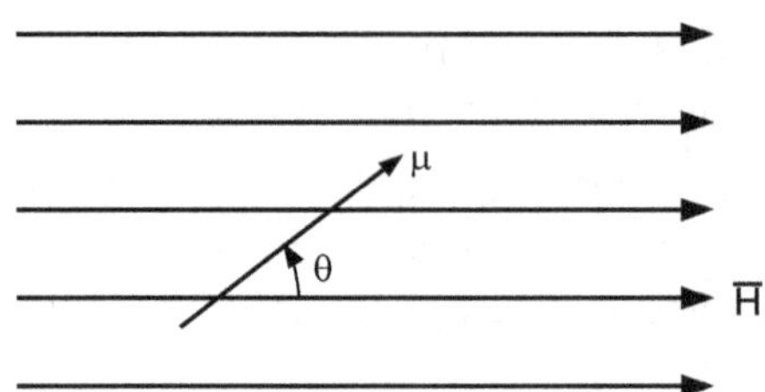

Fig. 4.5 : A magnetic dipole of moment $\bar{\mu}$ at angle θ relative to magnetic field $\bar{H}$

- A torque τ acting on a magnetic dipole in H is given by

$$\tau = \mu H \sin \theta \qquad \ldots (4.4)$$

where θ is the angle between $\bar{\mu}$ and H.

- τ is minimum when dipole is perpendicular to the field. τ is zero when it is parallel or antiparallel to the field. We have to calculate potential energy δE. To do this, we establish a reference configuration in which $\delta E = 0$.

- P.E. at any other orientation of $\bar{\mu}$ is the external work that must be done to rotate the dipole from $\theta_0 = 90°$ to the angle θ.

$$\therefore \quad \delta E = \int_{90°}^{\theta} \tau \, d\theta = \mu H \int_{90°}^{\theta} \sin \theta \, d\theta$$

$$\delta E = -\mu H \cos \theta \qquad \ldots (4.5)$$

When $\bar{\mu}$ points in the same direction of H, then

$$\delta E = -\mu H \text{ (minimum value)}$$

Magnetic Moment of Current Loop :

- In the magnetic moment of a current loop of area $\vec{A}$, through which current I is flowing, is given by

$$\bar{\mu} = I\vec{A} \qquad \ldots (4.6)$$

$$\mu = IA \text{ (considering only magnitude)}$$

- We know that an electron revolving in a circular orbit of radius r and making n revolutions is equivalent to current $(-ne)$ in a circular loop of area πr^2.

  The magnetic moment of electron will be

$$\bar{\mu} = -ne\pi r^2 \qquad \ldots (4.7)$$

- An electron of mass m, moving in a circular orbit of radius r with velocity v, has angular momentum

$$\vec{L} = m\vec{v} \, r$$

The velocity of electron will be

$$v = r\omega$$

$\therefore$

$$v = 2\pi nr$$

$\therefore$ The magnitude of angular momentum is

$$L = 2\pi mnr^2 \qquad \qquad \dots (4.8)$$

and

$$r^2 = \frac{L}{2\pi mn}$$

Substituting equation (4.8) in equation (4.7), we get

$$\mu = -\left(\frac{e}{2m}\right) L$$

where μ is the magnetic moment of electron due to orbital motion.
Let it be denoted by μ_l.

$\therefore$

$$\overline{\mu}_l = -\left(\frac{e}{2m}\right) \overrightarrow{L} \qquad \qquad \dots (4.9)$$

Substituting equation (4.9) in equation (4.5), we get

$$\delta E = \frac{e}{2m} LH \cos\theta \qquad \qquad \dots (4.10)$$

But

$$L = \sqrt{l(l+1)}\, \hbar = \sqrt{l(l+1)} \cdot \frac{h}{2\pi}$$

where, l is the orbital quantum number, h is the Planck's constant and

$$\cos\theta = \frac{L_z}{L} = \frac{m_l}{\sqrt{l(l+1)}}$$

where L_z is the component of angular momentum $\overrightarrow{L}$ in the direction of magnetic field vector $\overline{H}$ and m_l is the magnetic quantum number which has $(2l + 1)$ values from $- l$ to $+ l$ through zero.

Substituting value of L and $\cos\theta$ in equation (4.10),

$$\delta E = m_l \left(\frac{eh}{4\pi m}\right) H$$

$$= m_l \left(\frac{e\hbar}{2\pi m}\right) H$$

- The quantity $\left(\dfrac{eh}{4\pi m}\right) = \left(\dfrac{e\hbar}{2\pi m}\right)$ is called Bohr magneton and has a value 9.27×10^{-24} joule per tesla. Bohr magneton is unit of magnetic moment.

- Suppose an electron in atom goes from initial higher energy level E_{oi} to final lower energy level E_{of} when no magnetic field is supplied, then,

$$v_o = \frac{E_{oi} - E_{of}}{h}$$

- When magnetic field H is applied, the energy of initial and final states is given by

$$E_i = E_{oi} + m_{li}\left(\frac{eh}{4\pi m}\right)H$$

$$E_f = E_{of} + m_{lf}\left(\frac{eh}{4\pi m}\right)H$$

- The frequency of emitted photon (spectral line) is given by

$$v = \frac{E_i - E_f}{h} = \frac{E_{oi} - E_{of}}{h} + (m_{li} - m_{lf})\left(\frac{e}{4\pi m}\right)H$$

$$= v_o + \Delta m_l\left(\frac{e}{4\pi m}\right)H \qquad \qquad ...(4.11)$$

Selection rule for $\Delta m_l = +1, 0$ and -1.
Then we get three values of v given by

$$v_1 = v_o + \left(\frac{e}{4\pi m}\right)H = v_o + \Delta v$$

$$v_2 = v_o \qquad \qquad ...(4.12)$$

and
$$v_3 = v_o - \left(\frac{e}{4\pi m}\right)H = v_o - \Delta v$$

where
$$\Delta v = \left(\frac{e}{4\pi m}\right)H$$

But,
$$v = \frac{c}{\lambda}$$

$\therefore$
$$dv = -c\frac{d\lambda}{\lambda^2}$$

or
$$d\lambda = -\frac{\lambda^2}{c}dv$$

$\therefore$ Shift in wavelength is

$$d\lambda = \Delta\lambda = \pm\frac{\lambda^2}{c}\left(\frac{e}{4\pi m}\right)H \qquad \qquad ...(4.13)$$

4.4 Normal Zeeman Effect for Single Valence Electron System
(April 16)

- The effect of magnetic field on the spectrum of an atom, was first studied by Prof. Zeeman in the year 1896, when he kept a sodium flame between the pole-pieces of a strong electromagnet. He found that primarily a single spectral line splits up into 3 components - one line having larger frequency, other having lower frequency than the frequency of original line and third one having *frequency* of original line. Such splitting effect is named as normal Zeeman effect.

- Here as electron has two motions (orbital and spin), but due to application of strong magnetic field, l-s coupling gets broken and splitting is due to $\bar{l}$ vector only.

- If we neglect the spin of electron, then angular momentum possessed by electron is given as

$$p_l = l\frac{h}{2\pi}$$

- We know that orbital magnetic moment is given by

$$\mu_l = e\frac{lh}{4\pi m} = \frac{e}{2m}\,p_l$$

- Now in presence of an external magnetic field, the l vector precesses around the field direction. Such precession is called a Larmour precession.

  The precessional frequency is given as,

$$\omega_l = H\frac{\mu_l}{p_l} = \frac{e}{2m}\,H \qquad \qquad \text{... (4.14)}$$

(H is taken as magnetic field and $\dfrac{\mu_l}{p_l}$ ratio is known as gyromagnetic ratio.)

The electron gains an additional energy due to this precession, which can be written as :

$$\Delta E = \omega_l \times \text{Projection of orbital angular momentum on the field direction}$$

$$= \frac{e}{2m}\,H \cdot p_l \cos\theta$$

$$= \frac{eH}{2m}\frac{lh}{2\pi}\cos\theta$$

$$= \frac{eHh}{4\pi m}\,m_l \text{ where } m_l = l\cos\theta$$

where m_l takes $(2l + 1)$ values from $-l, ..., 0, ..., +l$. Hence a magnetic field splits up each energy level into $(2l + 1)$ sublevels, separation being $\dfrac{eh}{4\pi m}$.

- If E_{1H} and E_{2H} are the energies of two levels in presence of magnetic field and if E_1 and E_2 being energies in absence of magnetic field, with m_l values being m_{l1} and m_{l2} then

$$E_{1H} = E_1 + m_{l1}\frac{eHh}{4\pi m} \qquad \qquad \text{... (4.15)}$$

and

$$E_{2H} = E_2 + m_{l2}\frac{eHh}{4\pi m} \qquad \qquad \text{... (4.16)}$$

- Hence a radiation will be emitted in presence of a magnetic field if

$$E_{1H} - E_{2H} = (E_1 - E_2) + \frac{eHh}{4\pi m}(m_{l1} - m_{l2})$$

or
$$h\upsilon = h\upsilon_0 + \frac{eHh}{4\pi m}\,\Delta m_l$$

or
$$\upsilon = \upsilon_0 + \frac{eH}{4\pi m}\,\Delta m_l$$

where υ_0 is the frequency of a line in absence of magnetic field. The splitting will be observed, which is based on selection rule, $\Delta m_l = 0$ or ± 1.

Hence we observe three possible lines.

(i) $\upsilon_1 = \upsilon_0$ for $\Delta m_l = 0$ ... (4.17 a)

(ii) $\upsilon_2 = \upsilon_0 + \dfrac{eH}{4\pi m}$ for $\Delta m_l = +1$ ... (4.17 b)

(iii) $\upsilon_3 = \upsilon_0 - \dfrac{eH}{4\pi m}$ for $\Delta m_l = -1$ ... (4.17 c)

- In a magnetic field, the energy of a particular atomic state, depends on the value of m_l as well as on n.

- Therefore, a state of total quantum number n breaks up into various sub states, When an atom is kept in a magnetic field, and their energies are slightly more or less than energy of state in absence of magnetic field. This leads to splitting of individual spectral lines into separate lines when atoms radiate in a magnetic field.

- The spacing between the lines will depend on magnitude of the field. Fig. 4.6 shows normal Zeeman effect.

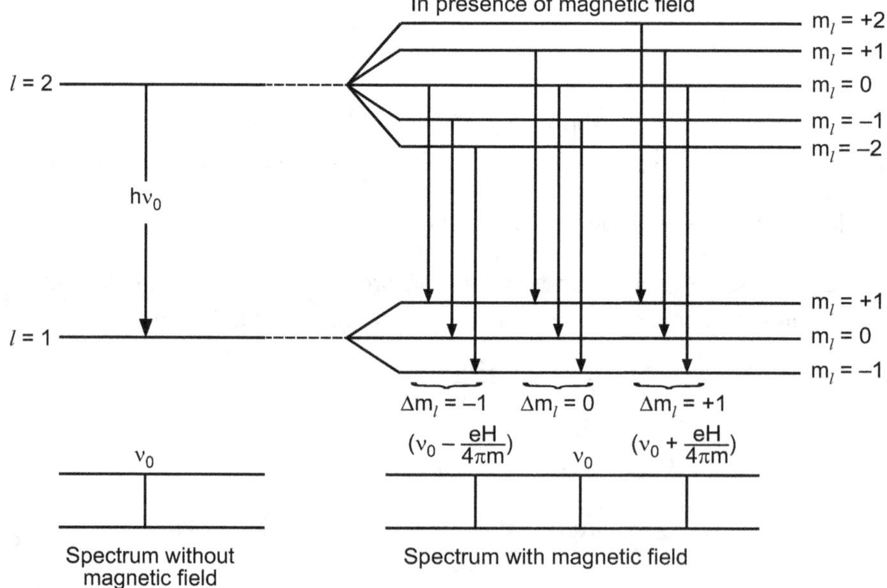

Fig. 4.6 : Normal Zeeman effect

- In the diagram, the three transitions in a bracket represent the same change in the value of Δm_l, hence same energy change and hence a single spectral line. It is to be remembered that Zeeman effect gives confirmation of space quantization.

4.5 Stark Effect (Qualitative Discussion)   (April 17, Oct. 15)

- Soon after the discovery of the Zeeman effect in 1896, a similar resolution of spectral lines in an applied electric field was observed. The action of an electric field on the spectrum of hydrogen was discovered by Stark in 1913, who observed the splitting of Balmer lines. This phenomenon is exhibited by the spectra of all elements and is known as Stark effect.

- It is rather surprising that the influence of an electric field on the spectral lines was discovered many years after the Zeeman effect. The main practical difficulty with the earlier investigators was the absence of a proper techniques. The main aim was to subject hydrogen atoms emitting spectral lines to a powerful electric field. This was not possible with ordinary Geissler tube containing hydrogen since the gas in such a tube is comparatively a good conductor and hence incapable of maintaining a strong electric field.

- Stark overcame this difficulty by placing an auxiliary electrode F close behind the cathode C at distance of few millimetres which is perforated as shown in Fig. 4.7. A very strong electric field of several thousand volts per cm is maintained between F and C. Positive ions are accelerated towards C and some will pick up electrons on the way forming neutral M atoms. The fast atoms (ions also) which pass through the perforations and emit light in FC are called canal rays. These canal rays were made to pass into the space between the plates of highly charged condenser.

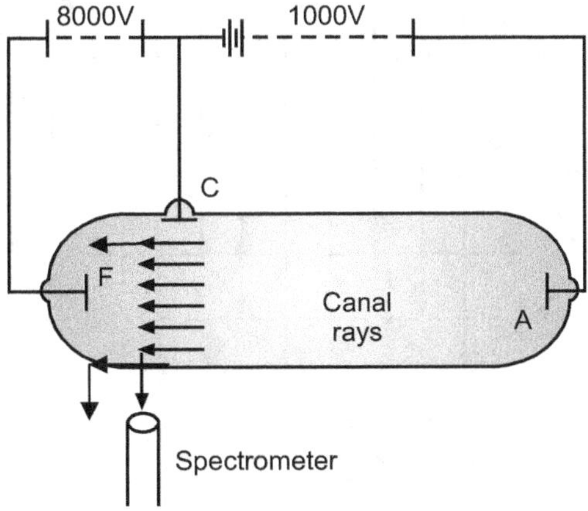

Fig. 4.7 : Stark's apparatus

- The main features of the phenomenon are the following:

  1. All hydrogen lines form symmetrical patterns, but the pattern depends markedly on the quantum number n of the term involved, the number of lines and the total width of the pattern increases with n. Thus, the number of components of H_β lines is greater than those of H_α line; similarly, the number of components of H_γ is greater than those of H_β.

  2. The wave number differences are integral multiples of a unit which is proportional to the field strength F and is the same for all hydrogen lines.

  3. Observation perpendicular to the direction of the electric field shows that the components are polarised in part parallel to the field (π components) and, in part, perpendicular to the field (σ components).

  4. Upto a field of about 1,00,000 volts per cm, the resolution increases in proportion to the field strength. In this region , we have linear Stark effect. In case of more intense field, more complicated effects, so called quadratic Stark effect and even of higher order are observed in addition to the linear effect.

- The energy level diagram of linear Stark effect in H-atom for n = 2 is shown in Fig. 4.8 in classical notations.

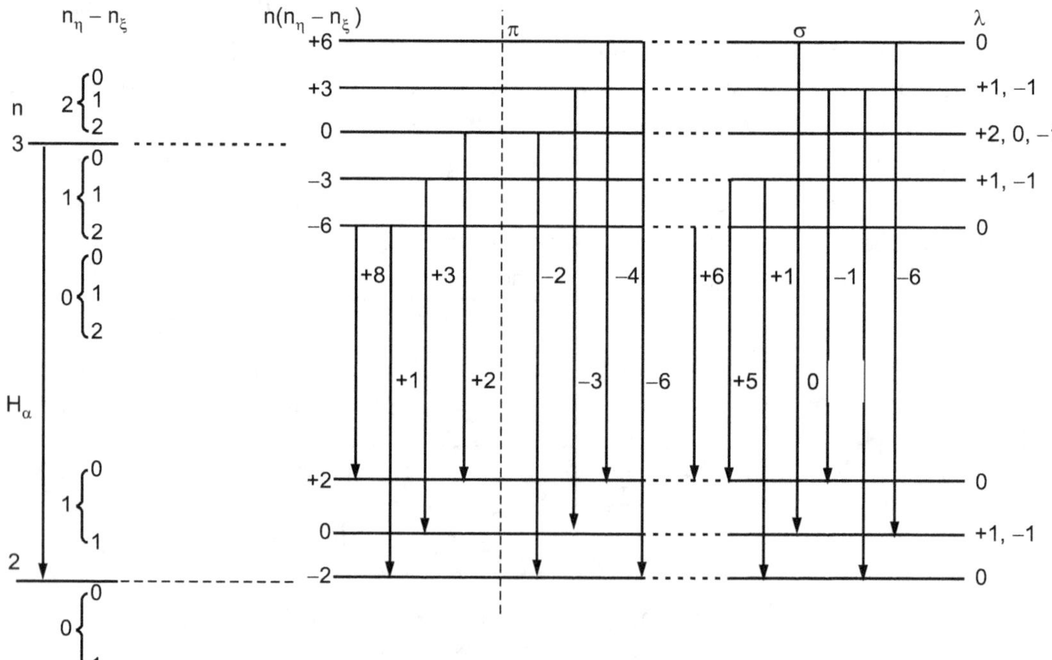

Fig. 4.8 : Stark pattern of the H_α line

- The discovery of this effect contributed importantly to the development of quantum theory. While first-order perturbation effects for the Stark effect in hydrogen are in agreement for the Bohr-Sommerfeld model and the quantum-mechanical theory of the atom, higher order effects are not. Measurements of the Stark effect under high field strengths confirmed the correctness of the quantum theory over the Bohr model. For ground state there is no first order Stark effect in case of hydrogen atom.

- The Stark effect is responsible for pressure broadening (Stark broadening) of spectral lines by charged particles. When splitted lines appear in absorption, the effect is called inverse Stark effect. The Stark effect can be explained with fully quantum mechanical approach.

- In a semiconductor heterostructure, where a small bandgap material is sandwiched between two layers of a larger bandgap material, the Stark effect can be dramatically enhanced by bound excitons. This is because the electron and hole which form the exciton are pulled in opposite directions by the applied electric field, but they remain confined in the smaller bandgap material, so the exciton is not merely pulled apart by the field. The quantum-confined Stark effect is widely used for semiconductor-based optical modulators, particularly for optical fiber communications.

- The Stark effect has been of marginal benefit in the analysis of atomic spectra but has been a major tool for molecular rotational spectra.

Solved Problems

Problem 4.1 : *Find the minimum magnetic field needed for Zeeman effect to be observed in a spectral line of 400 nm wavelength, when a spectrometer, whose resolution is 0.010 nm, is used.*

Solution : Given : $\lambda = 400 \text{ nm} = 400 \times 10^{-9} \text{ m}$

Here $\lambda = 0.010 \text{ nm} = 0.010 \times 10^{-9} \text{ m}$

Now, as $\upsilon = \dfrac{c}{\lambda}$

$\therefore$ $d\upsilon = -\dfrac{c}{\lambda^2} d\lambda$

Ignoring the negative sign,

$$d\upsilon = \Delta\upsilon = \frac{c}{\lambda^2} \Delta\lambda$$

$$= \frac{3 \times 10^8}{(400 \times 10^{-9})^2} \times (0.01 \times 10^{-9})$$

$$= \frac{30}{16} \times 10^{10} = 1.87 \times 10^{10} \text{ cycles/sec}$$

As the separation of Zeeman component is

$$\Delta \upsilon \ = \ \frac{eH}{4\pi m}$$

$\therefore \qquad H \ = \ \dfrac{4\pi m \, \Delta \upsilon}{e}$

$\therefore \qquad H \ = \ \dfrac{4 \times 3.14 \times 9.11 \times 10^{-31} \times 1.87 \times 10^{10}}{1.6 \times 10^{-19}}$

$\qquad\qquad = \ 133.7 \times 10^{-2}$

$\qquad\qquad = \ \textbf{1.337 Wb/m}^2 \qquad\qquad\qquad\qquad$ **... Ans.**

Problem 4.2 : *A sample of a certain element is placed in a 1 Tesla magnetic field and suitably excited. How far apart are the Zeeman components of the 5000 A° spectral line of this element ?*   **(April 17)**

Solution : Given : H = 1 Tesla, λ = 5000 A° = 5000×10^{-10} m, $\Delta\lambda$ = ?

The separation of the Zeeman components is

$$\Delta \upsilon \ = \ \frac{eH}{4\pi m}$$

Since, $\qquad\qquad \upsilon \ = \ \dfrac{c}{\lambda}$

$\therefore \qquad\qquad \Delta \upsilon \ = \ -\dfrac{c}{\lambda^2}\,\Delta\lambda$

(Here ignoring the negative sign)

$\therefore \qquad \Delta\lambda \ = \ \dfrac{\lambda^2}{c}\,\Delta\upsilon \ = \ \dfrac{\lambda^2}{c} \cdot \dfrac{eH}{4\pi m}$

$\qquad\qquad = \ \dfrac{(5000 \times 10^{-10})^2 \times 1.6 \times 10^{-19} \times 1}{3 \times 10^8 \times 4 \times 3.14 \times 9.11 \times 10^{-31}}$

$\qquad\qquad = \ 0.1165 \times 10^{-10}$ m

$\qquad\qquad = \ \textbf{0.1165 A}^\circ \qquad\qquad\qquad\qquad$ **... Ans.**

Problem 4.3 : *The Zeeman components of a 500 nm spectral line are 0.0106 nm apart when the magnetic field is 0.40 Tesla. Find the ratio $\dfrac{e}{m}$ for the electron from this data.*

Solution : Given : λ = 500 nm = 500×10^{-9} m, $\Delta\lambda$ = 0.0106 nm = 0.0106×10^{-9} m,
$\qquad\qquad\qquad$ H = 0.40 Tesla.

Now, as $\qquad\qquad \upsilon \ = \ \dfrac{c}{\lambda}$

$\therefore \qquad\qquad d\upsilon \ = \ \Delta\upsilon \ = \ -\dfrac{c}{\lambda^2}\,\Delta\lambda$

(Ignoring the negative sign.)

$$\therefore \quad \Delta\upsilon = \frac{c}{\lambda^2}\Delta\lambda = \frac{3\times10^8\times(0.0106\times10^{-9})}{(500\times10^{-9})^2}$$

$$= 1\times10^{-7}\times10^{17}$$

$$= 1\times10^{10} \text{ cycles/sec.}$$

As separation of Zeeman components is

$$\Delta\upsilon = \frac{eH}{4\pi m}$$

$$\therefore \quad \frac{e}{m} = \frac{4\pi\,\Delta\upsilon}{H} = \frac{4\times3.14\times1\times10^{10}}{0.40}$$

$$= \mathbf{31.4\times10^{10}\ Hz/Tesla} \qquad\qquad \textbf{... Ans.}$$

Problem 4.4 : *A spectral line of wavelength 4500 A° when placed in a magnetic field of 10 Tesla is observed to be a normal Zeeman triplet. Calculate wavelength separation between composition of triplet.*

Solution : Separation between Zeeman components is given by

$$\Delta\lambda = \pm\frac{\lambda^2}{c}\left(\frac{e}{4\pi m}\right)H$$

$$= \pm\frac{4500\times10^{-10}\times4500\times10^{-10}}{3\times10^8}\left(\frac{1.6\times10^{-19}}{4\pi\times9.1\times10^{-31}}\right)\times10$$

$$= \pm\,0.943 \text{ A°}$$

Hence Zeeman triplet has wavelength

$$\lambda_1 = 4500 - \Delta\lambda$$

$$= 4500 - 0.943 = 4499.057 \text{ A°}$$

$$\lambda_2 = 4500 - 0 = 4500 \text{ A°}$$

$$\lambda_3 = 4500 + \Delta\lambda$$

$$= 4500 + 0.943 = \mathbf{4500.943\ A°} \qquad\qquad \textbf{... Ans.}$$

Problem 4.5 : *The Zeeman component of a 500 nm spectral lines are 0.0116 nm apart, when magnetic field is 1 T. Find the ratio e/m for the electron from this data.*

Solution :

$$\Delta\lambda = \pm\frac{\lambda^2}{c}\frac{e}{4\pi m}H$$

$$\frac{e}{m} = \frac{4\pi c}{\lambda^2 H}\Delta\lambda$$

$$\lambda = 500\times10^{-9} \text{ nm,}$$

$$\Delta\lambda = 0.0116\times10^{-9} \text{ nm,} \quad H = 1 \text{ T}$$

$$\frac{e}{m} = \frac{4\pi c}{\lambda^2 H} \Delta\lambda$$

$$= \frac{4\pi \times 3 \times 10^8}{(500 \times 10^{-9})^2 \times 1} \times 0.0116 \times 10^{-9}$$

$$= \mathbf{1.757 \times 10^{11} \ C/kg} \qquad \text{... Ans.}$$

Problem 4.6 : *A sample of certain element is placed in magnetic field of flux density 0.3 T. How far apart are the Zeeman component of wavelength 4500 A° ?*

Solution :
$$\Delta\lambda = \pm \frac{\lambda^2}{c} \frac{e}{4\pi m} H$$

$$= \frac{4500 \times 10^{-10} \times 4500 \times 10^{-10}}{3 \times 10^8} \left(\frac{1.6 \times 10^{-19}}{4\pi \times 9.1 \times 10^{-31}} \right) \times 0.3$$

$$= \mathbf{0.0283 \ A°} \qquad \text{... Ans.}$$

Problem 4.7 : *Find the precessional frequency of an electron orbit when placed in a magnetic field of 6 T.*

Solution : The frequency $\Delta\upsilon_l$ of Larmor precession is given by

$$\Delta\upsilon_l = \frac{eH}{4\pi m}$$

Here we have, $H = 6T = 6 \ Wb/m^2$, $e = 1.6 \times 10^{-19} \ C$ and $m = 9.1 \times 10^{-31} \ kg$

$$\Delta\upsilon_l = \frac{1.6 \times 10^{-19} \times 6}{3 \times 3.14 \times 9.1 \times 10^{-31}}$$

$$= \mathbf{8.4 \times 10^{10} \ Hz} \qquad \text{... Ans.}$$

Summary

1. The splitting of spectral lines by a magnetic field is called Zeeman effect.

2. Classically, if υ_o is the orbital frequency of electron without field, the frequency in presence of field is given by $\upsilon_o \pm \Delta\upsilon$, where

$$\Delta\upsilon = \frac{eH}{4\pi mc}$$

3. From Lorentz classical treatment, Zeeman shift is given by

$$\Delta\upsilon = \frac{eH}{4\pi mc^2}$$

$$= 4.67 \times 10^{-5} \ H \ cm^{-1}$$

4. The splitting of spectral lines into three components in a magnetic field when viewed in a direction perpendicular to that of magnetic field is called normal Zeeman effect.

   Normal Zeeman effect is obtained from sources of elements like Ca, Cu, Zn, Cd, etc.

5. When a single spectral line is splitted into four or more lines when observed in a direction perpendicular that of magnetic field, it is known as anomalous Zeeman effect.

6. Change in wavelength in Zeeman effect is given by

$$d\lambda = \pm \frac{\lambda^2}{c}\frac{He}{4\pi m}$$

7. The Larmour precessional frequency is given as,

$$\omega_l = H\frac{\mu_l}{p_l} = \frac{e}{2m}H$$

$\dfrac{\mu_l}{p_l}$ ratio is known as gyromagnetic ratio

8. The Stark effect is the shifting and splitting of spectral lines of atom and molecules due to presence of an electric field. The amount of splitting or shifting is called Stark shift. The first order Stark effect is linear in applied electric field while the second order effect is quadratic in the field.

Exercise

(A) Short Answer Type Questions :

1. What is Zeeman effect ?
2. Define Bohr magneton.
3. What is anomalous Zeeman effect ?
4. Write formula for Zeeman shift.
5. Distinguish between normal Zeeman effect and anomalous Zeeman effect.
6. What is Stark effect?

(B) Long Answer Type Questions :

1. What is Zeeman effect ? Distinguish between normal Zeeman effect and anomalous Zeeman effect.
2. Explain in detail early discovery and developments in Zeeman effect.
3. With neat diagram explain experimental set-up to produce and observe Zeeman effect.
4. In Zeeman effect, show that

$$\upsilon = \upsilon_0 + \frac{eH}{4\pi m}\Delta m_l$$

   Hence discuss three cases.

5. Draw energy level diagram showing transitions in Zeeman effect. Hence write formulae for frequency for a spectrum with magnetic field.
6. Explain experimental set-up of Stark effect. Explain main features of Stark effect.

(C) Unsolved Problems :

1. The Zeeman components of a 650 nm spectral line are 0.014 nm apart, when magnetic field is 1.25 Tesla. Find the ratio e/m for the electron from this given data.

$$\left(\textbf{Ans. } \frac{e}{m} = 9.95 \times 10^{10} \text{ Hz/Tesla}\right)$$

2. The calcium line of wavelength λ = 4226.73 A° (P $\rightarrow$ S) exhibits normal Zeeman splitting when placed in uniform magnetic field of 2.5 Wb/m^2. Calculate the wavelength of three components of normal Zeeman pattern and separation between them.

   (**Ans.** λ_1 = 4226.5216 A°, λ_2 = 4226.73 A°, λ_3 = 4226.9384 A° and $\Delta\lambda$ = 0.2084 A°)

3. A sample of certain element is placed in 0.30 T magnetic field and suitably excited. How far apart the Zeeman component of the 450 nm spectral line of this element ?

   (**Ans.** 0.00283 nm)

4. The Zeeman components of a 500 nm spectral lines are 0.0116 nm apart when the magnetic field is one Tesla. Find the e/m for electron.

5. Determine the normal Zeeman effect of cadmium red line of 6438 A° when the atoms are placed in a magnetic field of 0.009 T.   (**Ans.** 1.74×10^{-3} A°)

6. Find the precessional frequency of an electron orbit when placed in a magnetic field of 5 T. Hence find the wavelength in normal Zeeman splitting.   (**Ans.** 6.998×10^{10} Hz)

Chapter **5**...

X-Ray Spectroscopy

Contents ...

Introduction

German Physicist - Roentgen

X-rays are electromagnetic radiations belonging to the electromagnetic spectrum lying between gamma radiations and ultraviolet radiations. The invention of X-rays is one of the milestones in the history of Physics. In 1895, Roentgen, a German Physicist, invented X-rays while working on discharge of electricity through gases. His simple observation opened up a series of discoveries in atomic physics, photochemistry and structure of material.

- X-rays are widely used in medicine (i.e. surgery and radiotherapy), engineering, industry and scientific research because of their special properties. In year 1912, M. Laue performed diffraction experiments and confirmed wave nature of X-rays. Now-a-days X-ray diffraction technique is an important tool in scientific research and Biotechnology.

- In this chapter, we discuss characteristic and continuous spectra, K, L, M, N series, comparison of X-ray with optical spectra and Moseley's law in detail. Fig. 5.1 is a diagram of an X-ray tube (Coolidge tube).

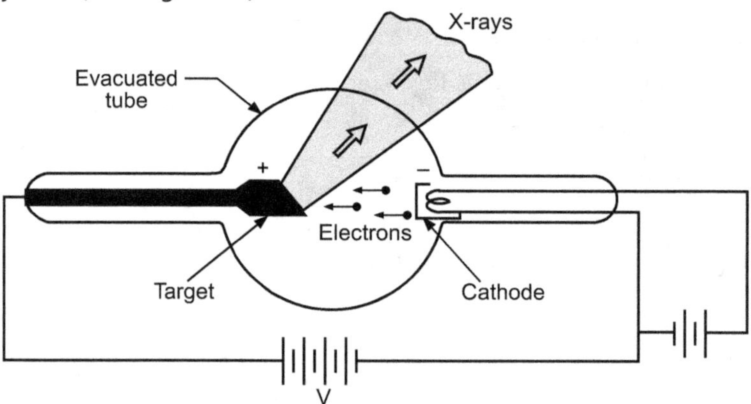

Fig. 5.1 : An X-ray tube

- A cathode, heated by a filament through which an electric current is passed, supplies electron by thermionic emission. A high potential difference V is maintained between the cathode and a metallic target. The electrons are attracted towards the cathode. The face of the target is at an angle relative to the electron beam. The X-rays that leave the target pass through the side of the tube. The tube is evacuated to permit the electrons to get to the target unimpeded.

- Fig. 5.2 shows a spectra when tungsten and molybdenum targets which are bombarded by electrons at several different accelerating potentials. The curve exhibits two features :

    1. In the case of molybdenum, intensity peaks occured indicate the enhanced production of X-rays at certain wavelength. These peaks occur at specific wavelength for each target material. These peaks are originated in rearrangement of electron structure of the target atom having been disturbed by bombarding electrons.

    2. The X-rays produced at a given accelerating potential V vary in wavelength but none has wavelength shorter than particular value of λ_{min}. As V increases, λ_{min} decreases.

    3. At a particular accelerating voltage, λ_{min} is same for tungsten and molybdenum targets.

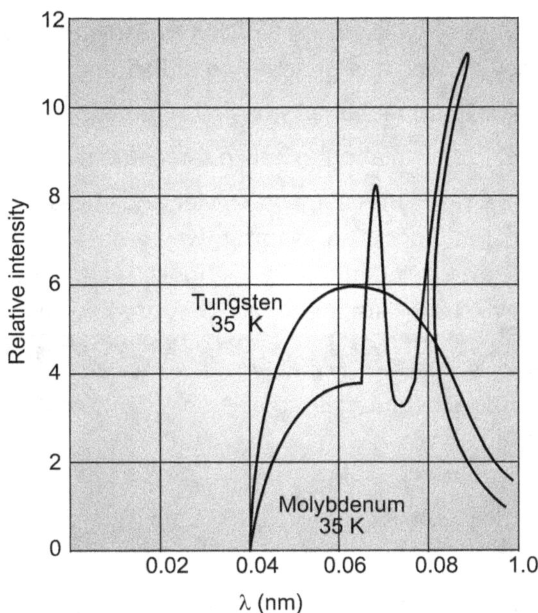

Fig. 5.2 : X-ray spectra of Tungsten and Molybdenum at 35 K

5.1 The Nature of X-Rays   (April 16, Oct. 15)

- Knowing that crystal is made up of regularly spaced atoms, Laue in 1912 suggested possibility of using a crystal as diffraction grating. When a narrow beam of X-ray is passed through crystal and then onto a photographic plate, a characteristic diffraction pattern is observed. The pattern composing chiefly of small symmetrically placed spots are observed. These spots are called Laue spots and entire pattern is known as Laue pattern. This experiment proved for the first time the wave nature of X-ray.

- Bragg Brother's (Bragg W.H. and Bragg W.L.) successfully showed X-ray diffraction by a crystal. It could be treated as reflection from crystal planes, the layers of regularly spaced atoms acting as reflecting surfaces. Knowing the distance d between successive layers of atoms in the crystal and equal angle of incidence and reflection θ, X-ray wavelength is given by

$$n\lambda = 2d \sin \theta$$

- Walter obtained single slit diffraction patterns of X-ray by sending these extremely short waves through a narrow slit and then onto a photographic plate. Interference fringes formed by the reflection of X-ray at a grazing angle from plane polarized mirrors have also seen and photographed.

- Bergen Davis and others have studied X-ray refraction and Comptonetal have diffracted X-ray from mechanically ruled grating.

Before wave nature of X-ray was known, Barkla and Sadler had observed the absorption of X-ray in aluminium. They showed that penetrating radiation consists of two groups of rays :

1. More penetrating or "hard radiation" is called K-radiation.

2. Less penetrating or "soft radiation" is called L-radiation.

- Photographs of XRD (X-ray diffraction) by crystal and measurement of λ showed that K-radiation characteristics consist of shorter wavelength and L radiation of longer wavelengths. From heavier element from periodic table soft radiation can be now observed and their wavelengths are measured. They are known as M, N and O radiation.

5.2 Duane and Hunt's Rule   (April 17; Oct. 17, 16, 11)

- Duane and Hunt found experimentally that *"short wavelength limit varies inversely as accelerating voltage"*.

$$\lambda_{min} \propto \frac{1}{V}$$

Their precise relation is

$$\lambda_{min} = \frac{hc}{eV}$$

$$= \frac{6.63 \times 10^{-34} \text{ Js} \times 3 \times 10^{8}}{1.6 \times 10^{-19} \text{ V}}$$

or $$\lambda_{min} = \frac{12400}{V} \text{ A}^{\circ} \qquad \qquad \text{... (5.1)}$$

This is called as *Duane and Hunt's relation.*

- The second observation i.e. shorter wavelength limit depends on accelerating voltage fits in quantum theory of radiation. Most of the electrons striking the targets undergo number of collisions, with their energy going into heat. (That is why the targets of X-ray tubes are made of materials having high melting point such as tungsten.)

- Few electrons, though loose most or all of their energy in single collision with the target atom, this is the energy that is evoked as X-rays.

- X-ray production, except for peaks mentioned above, represents inverse photoelectric effect. Instead of photon energy being transformed into electron kinetic energy (as in photoelectric effect), electron kinetic energy is transferred into photon energy. Since work functions are of the order of few eV, whereas the accelerating potentials in X-ray tube are typically of the order of kilovolt. We can ignore work function and interprete shorter wavelength limit as

$$eV = h\upsilon_{max} = \frac{hc}{\lambda_{min}}$$

$$\lambda_{min} = \frac{hc}{eV} = \frac{1.24 \times 10^{-6}}{V} \text{ volt-meters} \qquad \qquad \text{... (5.2)}$$

which is known as the Duane-Hunt's formula.

5.2.1 Origin of Continuous X-Ray Spectra   (Oct. 16)

- We have seen that X-ray spectra of target bombarded by fast electrons show narrow spikes at wavelength characteristics of the target material.
- These are in addition to a continuous distribution of wavelengths down to a minimum wavelength inversely proportional to electron energy. Continuous X-ray spectrum is the result of the inverse photoelectric effect.

Continuous X-Ray Spectrum :

- The X-rays emitted by target material have a continuous distribution of intensity above a certain minimum wavelength given by,

$$\lambda_{min} = \frac{1.242 \times 10^6}{V} = \frac{12420}{V} \; A^\circ$$

where V is the applied voltage in volt.

- The minimum wavelength is known as Duane-Hunt limit or continuous spectrum boundary. It is independent of nature of target.
- Continuous X-rays are produced by the phenomenon of Bremsstrahlung, which is a German word meaning 'braking' or 'slowing down' radiation. Electrons emitted from the cathode in the X-rays are accelerated towards the target, strike it, penetrate deep into the interior of its atoms and are attracted by the nuclei due to strong electrostatic interaction. The electron, therefore, deviates from its original path, as shown in Fig. 5.3.
- The deviation of the electron from its straight line path is equivalent to its collision with the nucleus. The electron loses more of its energy due to the collision. This energy appears as an X-ray photon. If E_i is the initial kinetic energy of the electron and E_f is the final kinetic energy, then the energy of the X-ray photon = $E_i - E_f$. If υ is the frequency of the X-ray photon emitted, then,

$$h\upsilon = E_i - E_f$$

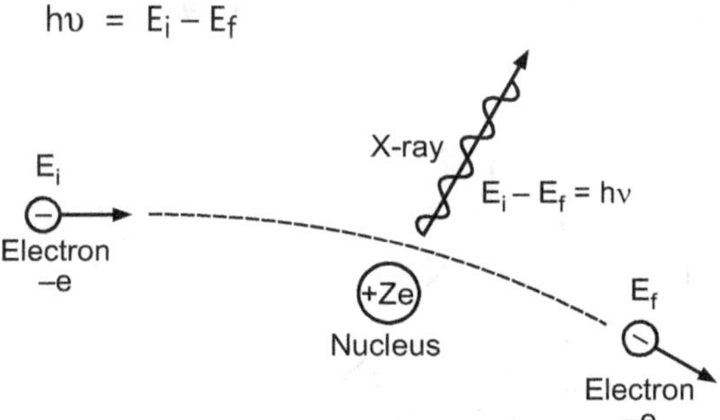

Fig. 5.3

- The electron may suffer several collisions with various nuclei before coming to rest and each collision is accompanied by the emission of an X-ray photon. Hence, a number of photons of different frequencies are emitted. As there are a large number of electrons in the beam and each electron suffers collision in a different way, we get photons of almost all frequencies or wavelengths, thus producing a continuous X-ray spectrum.

- *The production of continuous spectrum is the result of inverse photo-electric effect with electron kinetic energy $E_i - E_f$ being transformed into a photon energy $h\upsilon$.*

- According to classical theory, the continuous X-ray spectrum should consist of all frequencies between zero and infinity. But according to quantum theory, there is a minimum wavelength upto which the continuous X-ray spectrum extends. To find the value of λ_{min}, let the velocity of the incident electron decrease from v_i to v_f due to the interaction with the nucleus. Then frequency of X-ray photon is given by

$$h\upsilon = \frac{1}{2}mv_i^2 - \frac{1}{2}mv_f^2$$

- When electron is brought to rest by electrostatic interaction, $v_f = 0$ and photon has maximum frequency or minimum λ,

$$h\upsilon_{max} = \frac{1}{2}mv_i^2$$

If V is the P.D. between cathode and target then

$$\frac{1}{2}mv_i^2 = eV$$

$$h\upsilon_{max} = eV$$

$$\therefore \qquad \lambda_{min} = \frac{c}{\upsilon_{max}} = \frac{ch}{eV}$$

$$= \frac{1.242 \times 10^{-6}}{V}\ m$$

- The lowest wavelength limit of continuous spectra is inversely proportional to the accelerating potential of X-ray tube or $\lambda_{min} \propto 1/V$.

- The plot of relative intensity of X-ray spectra versus wavelength at various potential differences applied to an X-ray tube between the cathode and the target (Tungsten) is shown in Fig. 5.4.

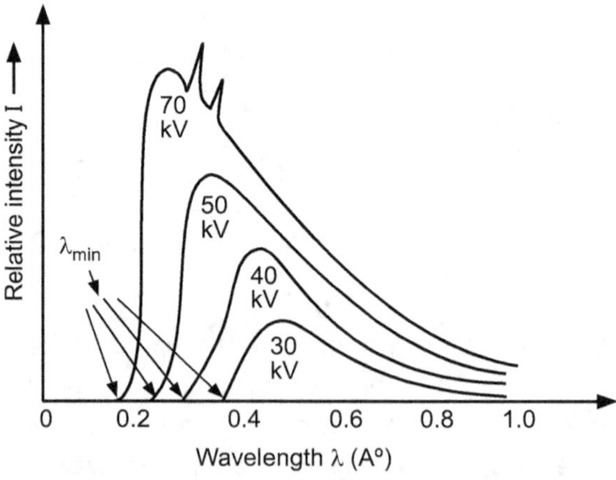

Fig. 5.4

The special features are :

(a) Change of target material : As $\lambda_{min} = \dfrac{ch}{eV}$, where c, h and e are constants, the cut-off wavelength is, therefore, the same for all target materials.

(b) Change in accelerating voltage :

(i)   The continuous spectrum ends abruptly at a minimum wavelength λ_{min}. The value of λ_{min} decreases with increasing accelerating potential.

(ii)   The intensity rises to a maximum at a certain wavelength, after which it gradually falls off. The intensity never reaches zero, showing that the radiation contains all possible wavelengths, above a certain minimum.

(iii) The wavelength corresponding to maximum intensity λ_m shifts gradually to lower wavelength side, as the accelerating potential is increased in such a way that $(\lambda_m \, V^{1/2})$ is almost a constant.

(iv)   For a particular target material, the intensity of continuous X-ray spectrum, corresponding to all wavelengths as well as the maximum intensity, increases with accelerating potential.

(v)   For applied potential difference 70 kV, two sharp peaks are seen. The sharp peaks are due to characteristic radiation. The line spectra is absent till the potential difference is greater than a particular value. The smoothly varying curves represent the continuous spectrum and the superimposed lines on the continuous background constitute the characteristic spectrum.

5.3 Characteristic X-Ray Spectra   (April 17, 11)

- The characteristic X-ray spectrum consists of sharp peaks superimposed on the continuous spectrum. Wavelength is characteristic of the target element. It is highly monochromatic and independent of the applied voltage. The intensity of the characteristic spectrum lines is very large - sometimes 10^4 times that of the continuous spectrum in that region.

- The characteristic wavelength is given by Moseley's law $\lambda \, (Z - b)^2 = a$ (constant) where b is different for different materials.

- Every substance gives rise to a particular set of characteristic X-rays, consisting of one or more series, known as K, L, M, etc. series. The X-rays in K series of a particular element have smaller wavelength than in the L or M series, whereas the wavelength of the X-rays in the same series say K, produced from an element of higher atomic weight is less than that produced from an element of lower atomic weight.

Production of Characteristic X-rays :   (Oct. 15)

- An atom consists of a central nucleus having a positive charge and revolving round it, in more or less circular orbits, a suitable number of electrons. The electron in the innermost orbit is attracted by the nucleus with the greatest force and to detach it from the atom, maximum energy is required. The innermost orbit is given the name K level or shell and the outer orbits L, M, etc. levels respectively. Thus more energy is required to detach an electron from an inner level than from an outer level.

- When an atom is bombarded by fast moving electrons, an electron coming with a high velocity penetrates through the outer shell containing electrons and removes an electron from one of the inner shells. An incoming electron knocking out an electron from K shell of an atom is shown in Fig. 5.5 (b). The incident and the dislodged electrons both rush out of the atom, causing a vacancy in the K shell or level.

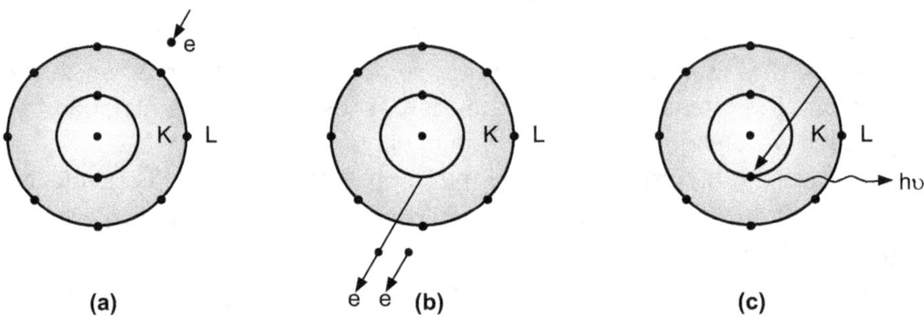

Fig. 5.5

- Immediately, one of the electrons from the higher levels (e.g. L level) jumps into the K level. A radiation of frequency υ is emitted at the same time. The frequency of the radiation is given by the relation,

$$E_K - E_L = h\upsilon$$

or
$$\upsilon = \frac{E_K - E_L}{h}$$

where E_K is the energy required to extract an electron from the K level, E_L is the energy required to extract an electron from the L level and h is Planck's constant having a value equal to 6.62×10^{-34} joule-sec. The frequency of the radiation is very high and is of the order of 10^{18} per second.

- The wavelength of the corresponding X-rays is given by,

$$\lambda = \frac{c}{\upsilon} = \frac{hc}{E_K - E_L}$$

- If an electron from the M level falls into the K level, X-rays of still higher frequency (shorter wavelength) will be produced. The frequencies of X-rays produced by the fall of an electron from an outer level to the K level can be arranged into a series, called K series, consisting of K_α, K_β, K_γ etc. lines, as shown in Fig. 5.6.

- A jump from L level to K level gives rise to K_α line and from M level to K level K_β line and so on, till we get limiting line, known as K-series limit.

- Similarly, when the incident electrons carry a lesser amount of energy, an electron is displaced from the L level and an electron from M or other outer level takes its place, so that a radiation of a lower frequency than that of K series is emitted. This corresponds to the L series of the X-ray spectrum. For the same element, the L lines of the X-ray spectrum have smaller frequency (larger wavelength) than the K lines. A jump from M shell to L shell gives rise to L_α line and from N shell to L shell gives rise to L_β line and so on, till we get the limiting line, known as L-series limit.

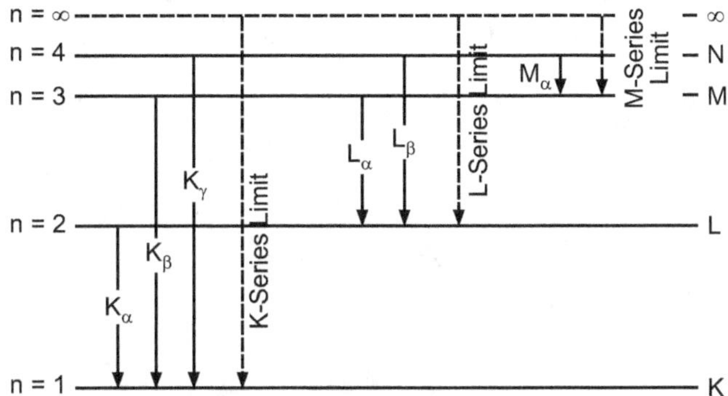

Fig. 5.6 : Spectral lines in different series and their transition

- The wavelengths of these lines in the K, L, or M series depend upon the atomic weight of the anti-cathode and these wavelengths are known as *characteristic X-rays* for the material of the given target.

5.4 Moseley's Law   (April 17, 11, 12; Oct. 17, 16, 15, 11)

- In 1913-14, Moseley carried out a systematic study of the characteristic X-ray spectra of various elements used as targets in an X-ray tube. For this purpose, he used Bragg's spectrometer to find that the characteristic X-ray spectra of different elements. They are remarkably similar to each other in the sense that each consists of K, L and M series. However, he found one very important difference. The frequency of lines in every series, produced from an element of higher atomic number, is greater than that produced by an element of lower atomic number. This is because the binding energy of electrons increases as we go towards the elements of higher atomic number. Since there is a greater positive charge on the nucleus of higher atomic number, large amount of energy is required to liberate an electron from the K, L and M shells of that element.

- Consider K_α line of the characteristic X-ray spectrum of any element. It is found that higher the atomic number of the target material, higher is the frequency of the K_α line produced by it. The exact mathematical equation between frequency and atomic number is given by,

$$\upsilon \propto (Z - b)^2$$
$$\sqrt{\upsilon} \propto (Z - b)$$
or $$\sqrt{\upsilon} = a\,(Z - b)$$

where Z is the atomic number of the element and a and b are constants for a particular series, but vary from one series to another. The constant b is known as 'nuclear screening constant'. For lines in K series, b = 1; for lines in L series, it's value is more. The relation is known as Moseley's law for the characteristic or line X-ray spectrum.

Statement : The frequency of a spectral line in the characteristic X-ray spectrum varies directly as the square of the atomic number of the element emitting it.

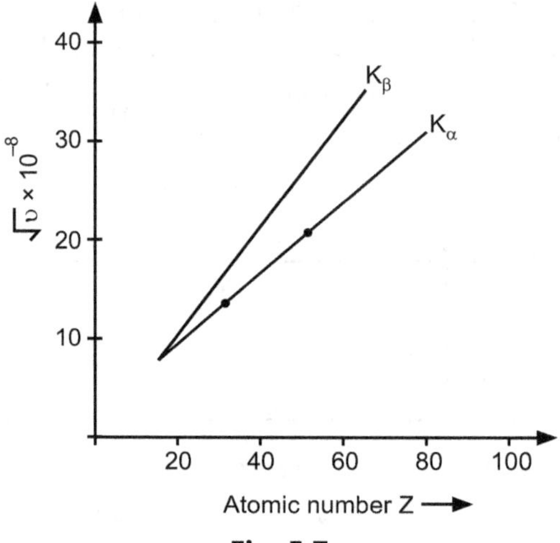

Fig. 5.7

- Fig. 5.7 shows Moseley's diagram for K_α and K_β lines, which is obtained by plotting $\sqrt{\upsilon}$ versus atomic number of different elements of the periodic table. The graph is linear as expected.

- The exact form of Moseley's law is

$$\bar{\upsilon} = \frac{1}{\lambda} = R\,(Z - \sigma^2)\left(\frac{1}{n_1^2} - \frac{1}{n_2^2}\right)$$

where R is the Rydberg's constant, Z is the atomic number, σ is the correction factor and n_1 and n_2 are the principal quantum numbers of the energy levels between which the transition occurs.

Importance of Moseley's Law :　　　　　　　　　　　　(Dec. 11)

- The great importance of Moseley's law lies in the fact that it proves for the first time that it is the atomic number and not the atomic weight of an element, which determines its characteristic properties (both physical and chemical). It gives the indication that the elements in the periodic table must be arranged according to their atomic numbers and not by their atomic weights. This fact is further supported by the chemical properties of the elements.

- Moseley's law has led to the discovery of new elements like hafnium (72), promethium (61), technetium (43) and rhenium (75) etc., by the indication of gaps in Moseley's diagram. The law has also helped in determining the atomic number of rare earth elements, thereby fixing their positions in the periodic table.

- Moseley's law is in accordance with Bohr's theory. When an electron jumps from an orbit n_2 to the orbit n_1, the frequency of radiation is given by,

$$\upsilon = \frac{me^4}{8\varepsilon_0^2 h^3} \cdot Z^2 \left(\frac{1}{n_1^2} - \frac{1}{n_2^2}\right)$$

$$= \left[\left(\frac{me^4}{8\varepsilon_0^2 h^3}\right)\left(\frac{1}{n_1^2} - \frac{1}{n_2^2}\right)\right] Z^2$$

$$\therefore \qquad \upsilon \propto Z^2 \quad \text{or} \quad \sqrt{\upsilon} \propto Z.$$

- Moseley took into account the screening effect of electrons, which Bohr did not. Thus the expression becomes $\sqrt{\upsilon} \propto (Z - b)$.

5.5 Fine Structure of X-Ray Spectra (April 12, Oct. 15)

- Moseley's observations about characteristic X-ray spectra were made in 1913. He observed that, most of the lines in the X-ray spectrum were found to consist of two or more lines. For example, K_α line was found to be a doublet i.e. two very close lines for all elements beyond Sodium (At. No. 11). Similarly, K_β line also showed a complex structure.

- Thus each X-ray line is found to consist of two or more lines very close together. This is known as *fine structure of X-ray lines*.

- To explain the emission of characteristic X-ray spectra, we considered the total energy difference only due to change of principal quantum number n. But to have a complete picture, the effect of changes in other quantum numbers have also to be taken into account. To a first approximation, the energy of an electron is determined by its configuration, which depends upon the principal as well as the orbital quantum number. Further, to take into account the sufficient large influence of spin quantum number, we take the vector sum of the orbital and spin quantum number, j – the total angular momentum quantum number.

- In the K-shell (n = 1, l = 0), the orbital is spherically symmetrical and only one value of j is possible i.e. $j = l + s = +\frac{1}{2}$. The number of levels in K-shell $(2l + 1) = 1$ i.e. there is only one level. In spectroscopic notation, it is denoted as $1s_{1/2}$.

- For the L-shell (n = 2, l = 1, 0), two values of j are possible, i.e. $j = \frac{3}{2}$ and $j = \frac{1}{2}$, corresponding to l = 1 i.e. $(l + s) = \left(1 + \frac{1}{2}\right) = \frac{3}{2}$ and $(l - s) = \left(1 - \frac{1}{2}\right) = \frac{1}{2}$ and one value $j = \frac{1}{2}$ corresponding to l = 0 i.e. $(l + s) = +\frac{1}{2}$. Thus there are three subshells L_I (n = 2, l = 0, j = $+\frac{1}{2}$), L_{II} (n = 2, l = 1 j = $\frac{1}{2}$) and L_{III} (n = 2, l = 1, j = $\frac{3}{2}$). In spectroscopic notations, these are denoted as $L_I = 2s_{1/2}$, $L_{II} = 2p_{1/2}$, and $L_{III} = 2p_{3/2}$.

- For the M-shell ($n = 3$, $l = 2, 1, 0$), there are five subshells, M_I ($n = 3$, $l = 0$, $j = \frac{1}{2}$) = $3s_{1/2}$,

  M_{II} ($n = 3$, $l = 1$, $j = \frac{1}{2}$) = $3p_{1/2}$, M_{III} ($n = 3$, $l = 1$, $j = \frac{3}{2}$) = $3p_{3/2}$, M_{IV} ($n = 3$, $l = 2$, $j = \frac{3}{2}$) = $3d_{3/2}$

  and M_V ($n = 3$, $l = 2$, $J = \frac{5}{2}$) = $3d_{5/2}$.

- Similarly, for N-shell there are seven subshells denoted as N_I = $4s_{1/2}$, N_{II} = $4p_{1/2}$, N_{III} = $4p_{3/2}$, N_{IV} = $4d_{3/2}$, N_V = $4d_{5/2}$, N_{VI} = $4f_{5/2}$, N_{VII} = $4f_{7/2}$. It should be noted that each subshell can have a maximum of $(2j + 1)$ electrons.

- This is known as the multiplicity of the shell and arises from the fact that the total angular momentum vector can be oriented in $(2j + 1)$ directions, according to the rule of spatial quantization. Thus, the multiplicity of the K-shell $\left(j = \frac{1}{2}\right)$ = 2 for $L_I\left(j = \frac{1}{2}\right)$ = 2, $L_{II}\left(j = \frac{1}{2}\right)$ = 2, $L_{III}\left(j = \frac{3}{2}\right)$ = 4, $M_I\left(j = \frac{1}{2}\right)$ = 2, $M_{II}\left(j = \frac{1}{2}\right)$ = 2, $M_{III}\left(j = \frac{3}{2}\right)$ = 4, $M_{IV}\left(j = \frac{3}{2}\right)$ = 4, $M_V = \left(j = \frac{5}{2}\right)$ = 6 and so on for N subshells.

- An energy level diagram, based on the above data, is shown in Fig. 5.8, in which each level represents an allowed electron state. If an electron is removed from a sub-shell, the atom gets ionised. The vacancy thus created may be filled by an electron from an outer shell, according to selection rules, giving rise to fine structure of the lines. These selection rules are

$$\Delta l = \pm l; \quad \Delta j = 0, \pm 1, \quad \Delta n = \text{any value}$$

- K_α line, which is produced by the transition of an electron from L-level to K-level, will now consist of two lines K_α (1) and K_α (2) for transitions from L_{III} or $2P_{3/2}$ to K or $1s_{1/2}$ and from L_{II} or $2p_{1/2}$ to K or $1s_{1/2}$, level according to $l = 1 \to l = 0$ or $\Delta l = +1$ and $j = \frac{3}{2} \to j = \frac{1}{2}$ or $\Delta j = +1$ and $l = 1 \to l = 0$ or $\Delta l = +1$ and $j = \frac{1}{2} \to j = \frac{1}{2}$ or $\Delta j = 0$.

- The transition from L_I to K (or $2s_{1/2}$ to $1s_{1/2}$) involving $l = 0 \to l = 0$ or $\Delta l = 0$ is not allowed. Thus K_α line is a doublet. K_α and K_γ lines also consist of two very closed lines each. Some allowed transitions giving rise to fine structure in K_α, K_β and L_α lines are shown in Fig. 5.8.

- There are other weak lines observed in X-ray spectra which cannot be expressed as difference of any two levels given in Fig. 5.8. These lines are called non-diagram lines and generally lie close to one of the diagram lines. These are known as satellite lines and are produced due to transitions in atoms from which two or more electrons have been removed.

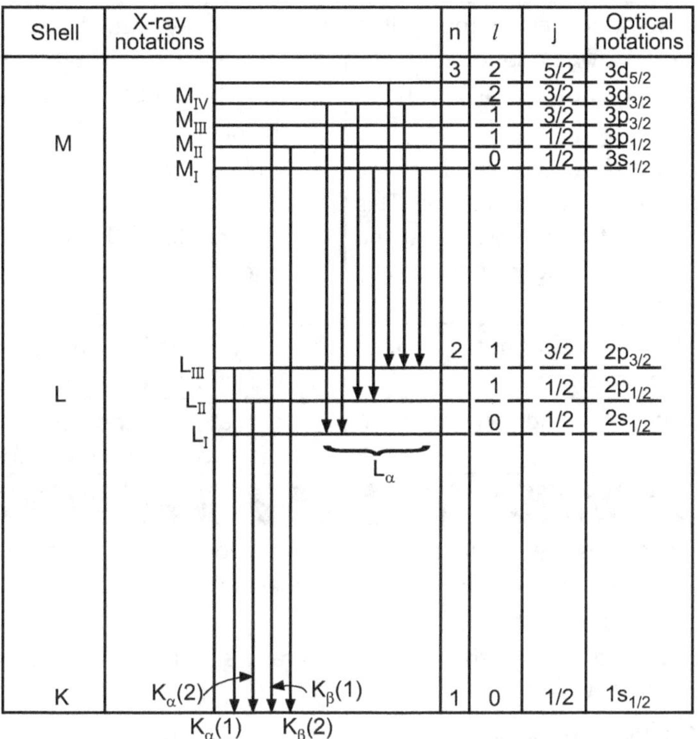

Fig. 5.8 : Fine structure of X-ray spectra

5.6 Comparison of X-Ray and Optical Spectra

- The X-ray spectra emitted by different elements are like one another and simpler than corresponding optical spectra. The main differences between the two spectra are as follows :

Optical Spectra	X-Ray spectra
1. This spectra is observed when there is a transition of electron located in the outermost shell of an atom. Such electron is also known as optical electron.	1. X-ray spectra is observed when there is a transition of electron from inner shells (such as K, L, M ... shells) of an atom.
2. Optical photon has less energy (or small frequency).	2. As energies of inner shells are much large, X-ray photon emitted has high energy (or high frequency)
3. Optical radiation has less penetrating power.	3. The X-ray radiation has very high penetrating power. (X-ray photon is 10^4 times more penetrating than an optical photon).

... Contd.

4.	The optical radiation lies in the visible range of the electromagnetic spectrum (wavelength range being 400 nm to 700 nm).	4.	The X-ray radiation lies in between gamma ray to ultra-violet region of electromagnetic spectrum. (Wavelength being 0.01 nm to 10 nm).
5.	Optical spectra show abrupt periodic changes from element to element. This is because the electronic structure at the surface of atom changes periodically from element to element.	5.	X-ray spectra vary smoothly from element to element because X-ray spectra depends on B.E. of electron in inner shells.
6.	Optical radiation does not produce ionization and fluorescence.	6.	X-ray radiation can produce ionization and fluorescence.
7.	Optical radiation is not harmful to human body.	7.	Repeated exposure to X-rays are hazardous to human body.

5.7 Applications of X-Rays

- Applications of X-rays are found to be in many fields such as medical, industrial, basic research, crystal structures, etc. Few of them are as follows:
  1. Detection of bone fractures.
  2. Determination of crystal structures using Bragg's diffraction condition $n\lambda = 2d \sin \theta$.
  3. Structure of DNA.
  4. X-ray imaging technique.
  5. Computed tomography (CT) technique for three dimensional images inside the body.

5.8 Auger Effect   (April 2011)

- In atoms, the filling of a vacancy in an inner electron energy level by an electron from an outer energy level of the same atom takes place. The excess energy involved causes emission of another electron known as an Auger electron. Auger electrons were discovered independently by Lise Meitner (in 1923) and the French physicist Pierre Victor Auger (in 1925), but the English-speaking scientific community attached Auger's name to effect.

- The Auger effect is a phenomenon in which the transition of an electron in an atom filling in an inner-shell vacancy causes the emission of another electron. When an electron is removed from a core level of an atom, leaving a vacancy, an electron from a higher energy level may fall into the vacancy, resulting in a release of energy. Although sometimes this energy is released in the form of an emitted photon, the energy can also be transferred to another electron, which is ejected from the atom. This second ejected electron is called an Auger electron.

- Upon ejection, the kinetic energy of the Auger electron corresponds to the difference between the energy of the initial electronic transition and the ionization energy for the electron shell from which the Auger electron was ejected. These energy levels depend on the type of atom and the chemical environment in which the atom was located.

- Auger electrons are produced when a sample is bombarded with electrons and a characteristic X-ray produced by inner shell ionization is reabsorbed, ejecting an electron. The Auger effect forms the basis of Auger spectroscopy.

- Auger spectroscopy measures the energy of electrons from the Auger effect of excited electrons in the surface region. There are two different ways that an atom with an excited core electron (a core electron hole) can return to its ground state. The first is that an energetically higher electron "jumps" to the hole left by the excited core electron, thereby releasing characteristic X-ray radiation. The second is the Auger effect, by which the hole left by the excited core electron is filled by an outer electron. The energy from this is transferred by a radiationless process to another electron that will then be ejected from the surface with a characteristic kinetic energy. The result is a spectrum of the intensity of electrons as a function of the kinetic energy of the Auger electron with the placement of the peaks giving a qualitative view of what elements are present on the surface.

- One of the two principal processes for the relaxation of an inner-shell electron vacancy in an excited or ionized atom. The Auger effect is a two-electron process in which an electron makes a discrete transition from a less bound shell to the vacant, but more tightly bound, electron shell. The energy gained in this process is transferred, via the electrostatic interaction, to another bound electron which then escapes from the atom. This outgoing electron is referred to as an Auger electron and is labelled by letters corresponding to the atomic shells involved in the process. For example, a KL_IL_{III} Auger electron corresponds to a process in which an L_I electron makes a transition to the K shell and the energy is transferred to an L_{III} electron (illus. a). By the conservation of energy, the Auger electron kinetic energy E is given by $E = E(K) - E(L_I) - E(L_{III})$ where E(K, L) is the binding energy of the various electron shells. Since the energy levels of atoms are discrete and well understood, the Auger energy is a signature of the emitting atom.

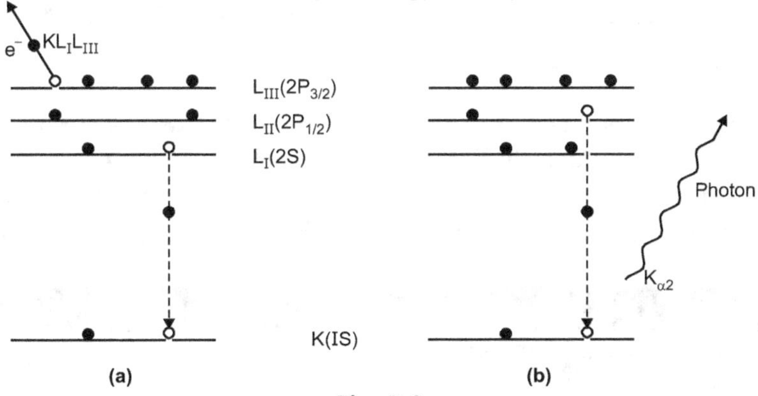

Fig. 5.9

- The other principal process for the filling of an inner-shell hole is a radiative one in which the transition energy is carried off by a photon (illus. b). Inner-shell vacancies in elements with large atomic number correspond to large transition energies and usually decay by such radiative processes; vacancies in elements with low atomic number or outer-shell vacancies with low transition energies decay primarily by Auger processes.

- Two principal processes for the filling of an inner-shell electron vacancy. (a) Auger emission; a $KL_I L_{III}$ Auger process in which an L_I electron fills the K-shell vacancy with the emission of a $KL_I L_{III}$ Auger electron from the L_{III} shell. (b) Photon emission; a radiative process in which an L_{II} electron fills the K-shell vacancy with the emission of a $K_{\alpha 2}$ photon.

- Auger electron spectroscopy (AES) has now emerged as one of the most widely used analytical techniques for obtaining the chemical composition of solid surfaces. The basic advantages of this technique are its high sensitivity for chemical analysis in the 5 to 20 A° region near the surface, a rapid data acquisition speed, its ability to detect all elements above helium, and its capability of high-spatial resolution. The high-spatial resolution is achieved because the specimen is excited by an electron beam that can be focussed into a fine probe. It was developed in 1960s, when ultra-high vacuum (UHV) technology became commercially available.

Or in short

Auger Effect

- When a K-electron is ejected from an atom say, by the absorption of an X-ray photon, the resulting vacancy in the K-shell is filled by the transition of an electron from an outer shell, for example, the L-shell which corresponds to a transition of the atom from the K-level to the L-level. This transition is usually accompanied by the emission of characteristic X-ray photon of energy.

$$h\upsilon' = E_K - E_L$$

- In some cases, however, the entire energy emitted in the transition is directly absorbed by another L-electron which is thereby ejected. Since there are now two vacancies in the L-shell, the resulting energy level of the atom will be different from the L-level and may now be called as LL-level, its energy being E_{LL}. The kinetic energy of the ejected second electron will be

$$\frac{1}{2}mv^2 = E_K - E_{LL}$$

- Such a raditionless transition resulting in the ejection of two electrons from the same atom is known as Auger transition. The phenomenon is known as Auger effect.

- The Auger effect is competitive with X-ray emission in most atoms, but the resulting ejected electrons are usually absorbed in the target material,from which the X-ray readily emerge.

Solved Problems

Problem 5.1 : *Calculate the minimum voltage that must be applied to an X-ray tube to produce X-ray photons of $\lambda \sim 0.1$ A°.*

(Given : $h = 6.63 \times 10^{-34}$ Js, $c = 3 \times 10^8$ m/s and $e = 1.6 \times 10^{-19}$ C).

Solution : When electrons accelerated through a potential V strike a target the maximum frequency υ_{max} (or minimum wavelength λ_{min}) of the emitted X-ray photon is given by

$$eV = h\upsilon_{max} = \frac{hc}{\lambda_{min}}$$

The minimum voltage for 0.1 A° wavelength of X-ray photon is

$$V = \frac{hc}{e\lambda_{min}}$$

$$= \frac{6.63 \times 10^{-34} \times 3 \times 10^{8}}{1.6 \times 10^{-19} \times 0.1 \times 10^{-10}}$$

$$= 1.24 \times 10^{5} \text{ J/C} = \mathbf{1.25 \times 10^{5} \text{ V}} \qquad \text{... Ans.}$$

Problem 5.2 : *Estimate the value of the wavelength of K line of silver (Z = 47).*
*(**Given :** R = 109737 cm^{-1})*

Solution : From Moseley's law, the wavelength of K line is given by

$$\frac{1}{\lambda} = R(Z-1)^{2}\left[\frac{1}{1^{2}} - \frac{1}{2^{2}}\right]$$

$$= \frac{3}{4} R(Z-1)^{2}$$

$$= \frac{3}{4} \times 109737 \text{ cm}^{-1} \times (46)^{2}$$

$$= 1.74 \times 10^{8} \text{ cm}^{-1}$$

$$\lambda = \mathbf{0.57 \text{ A}°} \qquad \text{... Ans.}$$

Problem 5.3 : *If the K radiation of Mo (Z = 42) has a wavelength of 0.71 A°, calculate the wavelength of the corresponding radiation of Cu (Z = 29).*

Solution : From Moseley's law, for K line, we have

$$\frac{1}{\lambda} = (Z-1)^{2}$$

$$\frac{\lambda_{Cu}}{\lambda_{Mo}} = \frac{(Z_{Mo}-1)^{2}}{(Z_{Cu}-1)^{2}}$$

$$= \frac{(41)^{2}}{(28)^{2}}$$

But

$$\lambda_{Mo} = 0.71 \text{ A}°$$

$$\lambda_{Cu} = 0.71 \times \frac{(41)^{2}}{(28)^{2}} = \mathbf{1.52 \text{ A}°} \qquad \text{... Ans.}$$

Problem 5.4 : *Find the nuclear screening constant for L series of X rays if it is known that X-rays with a wavelength of λ = 1.43 A° are emitted when an electron in a tungsten atom (Z = 74) is transferred from M level to L level. (Take R = 10.97 × 10⁶ m⁻¹).*

Solution : When electron jumps from M to L level, the first member of the L series i.e. L_α line is given out. Its wavelength by Moseley's law is

$$\frac{1}{\lambda} = R(Z-b)^{2}\left(\frac{1}{2^{2}} - \frac{1}{3^{2}}\right) = \frac{5R}{36}(Z-b)^{2}$$

Substituting

$$\frac{10^{10}}{1.43} = 10.97 \times 10^{6} \times \frac{5}{36}(74-b)^{2}$$

$$\therefore \qquad b = \mathbf{6.25} \qquad \text{... Ans.}$$

Problem 5.5 : *Find the shortest wavelength present in the radiation from an X-ray machine whose accelerating potential is 100 K.*   **(Dec. 11, April 12)**

Solution : Minimum wavelength is given by,

$$\lambda_{min} = \frac{ch}{eV}$$

$$= \frac{3 \times 10^8 \times 6.64 \times 10^{-34}}{1.6 \times 10^{-19} \times 100 \times 10^3}$$

$$= \frac{1.242 \times 10^{-6}}{100 \times 10^3}$$

$$= 1.242 \times 10^{-11}$$

$$= \textbf{0.0124 nm} \qquad \textbf{... Ans.}$$

Problem 5.6 : *Electrons are accelerated in a television tube through potential difference of 10 kV. Find the highest frequency and minimum wavelength of the electromagnetic waves emitted, when these strike the screen of the tube. In which region of the spectrum will these waves lie ?*

Solution : Minimum wavelength,

$$\lambda_{min} = \frac{ch}{eV} = \frac{1.242 \times 10^{-6}}{V}$$

Here, $V = 10$ kV

$$\therefore \qquad \lambda_{min} = \frac{1.242 \times 10^{-6}}{10 \times 10^3} = \textbf{1.242 A}° \qquad \textbf{... Ans.}$$

$$\upsilon_{max} = \frac{c}{\lambda_{min}} = \frac{3 \times 10^8}{1.242 \times 10^{-10}} = \textbf{2.42} \times \textbf{10}^{18} \textbf{ Hz} \qquad \textbf{... Ans.}$$

The waves lie in X-ray region of electromagnetic radiation (X-rays lie in wavelength region 0.3 A° - 100 A°).

Problem 5.7 : *If K, L and M energy levels of platinum are 80000, 14000 and 5000 eV respectively, calculate wavelength of K_α and K_β lines from platinum.*

Given : *1 eV = 1.6 × 10⁻¹⁹ J, h = 6.6 × 10⁻³⁴ Js.*

Solution : Energy of K level region,

$$E_K = 80000 \text{ eV}$$

$$= 80000 \times 1.6 \times 10^{-19} \text{ J}$$

Energy of L level, $E_L = 14000 \times 1.6 \times 10^{-19}$ J

Energy of M level, $E_M = 5000 \times 1.6 \times 10^{-19}$ J

Wavelength for K_α line,

$$\lambda_1 = \frac{hc}{E_K - E_L}$$

$$= \frac{6.6 \times 10^{-34} \times 3 \times 10^8}{(80000 - 14000) \times 1.6 \times 10^{-19}} = \frac{19.8 \times 10^{-26}}{66 \times 1.6 \times 10^{-16}}$$

$$\lambda_1 = \textbf{0.1875 A}° \qquad \textbf{... Ans.}$$

Similarly,

$$\text{Wavelength for } K_\beta \text{ line} = \frac{hc}{E_M - E_K}$$

$$= \frac{6.6 \times 10^{-34} \times 3 \times 10^8}{(80000 - 5000) \times 1.6 \times 10^{-19}}$$

$$= 0.165 \text{ A}° \qquad \text{... Ans.}$$

Summary

- X-rays are widely used in medicine (i.e. surgery and radiotherapy), engineering, industry and scientific research because of their special properties.

- Duane and Hunt found experimentally that "short wavelength limit varies inversely as accelerating voltage".

- Continuous X-ray spectrum is the result of the inverse photoelectric effect.

- The production of continuous spectrum is the result of inverse photo-electric effect with electron kinetic energy $E_i - E_f$ being transformed into a photon energy $h\upsilon$.

- The characteristic X-ray spectrum consists of sharp peaks superimposed on the continuous spectrum. Wavelength is characteristic of the target element.

- **Moseley's law :**
  Statement : The frequency of a spectral line in the characteristic X-ray spectrum varies directly as the square of the atomic number of the element emitting it.

- The great importance of Moseley's law lies in the fact that it proves for the first time that it is the atomic number and not the atomic weight of an element, which determines its characteristic properties (both physical and chemical).

- Moseley's law has led to the discovery of new elements like hafnium (72), promethium (61), technetium (43) and rhenium (75) etc., by the indication of gaps in Moseley's diagram.

- Each X-ray line is found to consist of two or more lines very close together. This is known as fine structure of X-ray lines.

- The X-ray spectra emitted by different elements are like one another and simpler than corresponding optical spectra.

- The Auger effect is a phenomenon in which the transition of an electron in an atom filling in an inner-shell vacancy causes the emission of another electron.

- Auger spectroscopy measures the energy of electrons from the Auger effect of excited electrons in the surface region.

- The Auger effect is a two-electron process in which an electron makes a discrete transition from a less bound shell to the vacant, but more tightly bound, electron shell.

- A KL_IL_{III} Auger electron corresponds to a process in which an L_I electron makes a transition to the K shell and the energy is transferred to an L_{III} electron (illus. a).

- The other principal process for the filling of an inner-shell hole is a radiative one in which the transition energy is carried off by a photon (illus. b).

- Such a raditionless transition resulting in the ejection of two electrons from the same atom is known as Auger transition. The phenomenon is known as Auger effect.

Exercise

(A) Short Answer Type Questions :

1. What are different types of spectra ?
2. State Duane and Hunt's law.
3. What is the importance of Moseley's law ?
4. How X-ray production is inverse of photoelectric effect ?

(B) Long Answer Type Questions :

1. What are X-rays ? How they are produced ?
2. State and explain Duane and Hunt's law.
3. Discuss origin of continuous X-ray spectrum.
4. What is the characteristic X-ray spectrum ? Discuss its origin.
5. State and explain Moseley's law. Discuss applications of Moseley's law.
6. Using energy level diagram, obtain the electron transition for K_{α_1}, K_{α_2} and K_{β_1}, K_{β_2} lines.
7. Compare X-ray spectra with optical spectra.

(C) Numerical Problems :

1. X-ray tube operating at 44 kV emits X-rays of wavelength 0.284 A°. Calculate Planck's constant. (**Ans.** 6.62×10^{-34} Js)

2. What voltage must be applied to an X-ray tube for it to emit X-rays with minimum wavelength 5000 A° ? (**Ans.** 24.85 kV)

3. Determine wavelength of K_α X-rays emitted by an element having Z = 79, a = 2.468×10^{15} s^{-1} and b = 1. (**Ans.** 0.1977 A°)

4. The effective charge experienced by an M (n = 3) electron in an atom of atomic number Z is about (Z – 7.4) e. Show that frequency of L_α X-rays of such element is given by $\dfrac{5CR\,(Z-7.4)^2}{36}$.

5. Calculate the maximum velocity of electrons striking the cathode under accelerating potential of 20 kV. Given : m = 9.11×10^{-31} kg, e = 1.602×10^{-19} C.

6. Find the wavelength for which first order Bragg's reflection occurs at a 10° glancing angle from calcite crystal.

Chapter **6**...

Molecular Spectroscopy

Contents ...

Introduction

- Molecular spectroscopy is the study of the interaction of electromagnetic waves and matter which gives us more information related to the structure of matter. Electromagnetic radiations can be considered as simple harmonic waves propagated from source. These electromagnetic waves are the characteristics of the respective source. Any matter consists of atoms and molecules. The atoms are bounded together to form molecule. In 1961, G.W. Lewis presented his theory of valency and explained the bonding. According to Pauding, a chemical bond exists between two atoms if the bonding force is stronger enough to have stability to consider these number of atoms as an independent molecular species. The electromagnetic radiations given by the molecule depend on the types of the atoms and the types of the bonding between them.

- In this chapter, we discuss rotational energy levels, vibrational energy levels and electronic spectra of molecule.

Molecular Spectra   **(April 11, 12)**

- There are many ways of classifying electromagnetic radiations given out by atoms and molecules. These can be classified by the frequency range. The regions do not have any fixed boundaries as such which can give us the precise wavelength. However, the molecular processes associated with each region are quite different, still we can have the classification depending on different regions as follows :

	Region	Frequency	Wavelength
1.	Radiofrequency region	$3 \times 10^6 - 3 \times 10^{10}$ Hz	10 m - 1 cm
2.	Microwave region	$3 \times 10^{10} - 3 \times 10^{12}$ Hz	1 cm - 100 μm
3.	Infrared region	$3 \times 10^{12} - 3 \times 10^{14}$ Hz	100 μm - 1 μm
4.	Visible & Ultraviolet region	$3 \times 10^{14} - 3 \times 10^{16}$ Hz	1 μm - 10 nm
5.	X-ray region	$3 \times 10^{16} - 3 \times 10^{18}$ Hz	10 nm - 100 pm
6.	γ-ray region	$3 \times 10^{18} - 3 \times 10^{20}$ Hz	100 pm - 1 pm

Types of Molecular Spectra :   **(April 17, 16; Dec. 11)**

- We know that molecules can remain in the ground state as well as excited states. The transition between the allowed energy states of a molecule with the emission or absorption of radiation give rise to molecular spectra. The frequency of emitted or absorbed photon is given by

$$\upsilon = \frac{E_2 - E_1}{h} \qquad \text{if} \quad E_2 > E_1 \text{ (i.e. absorption)}$$

or

$$\upsilon = \frac{E_1 - E_2}{h} \qquad \text{if} \quad E_1 > E_2 \text{ (i.e. emission)}$$

- The energy levels of molecule are more complex than atom. In case of molecule, various types of motion of individual nuclei can take place.

- The excited energy level of a diatomic molecule for a fixed electronic configuration arise from :

    (i)   rotation of molecule about an axis passing through the centre of mass of molecule and perpendicular to the line joining atoms.

    (ii) the vibrations of atoms about their equilibrium position in molecule.

    Therefore, total energy consists of :

    (i)   Energy due to excitation of electron in the molecule i.e. Electronic energy, E_e.

    (ii) Energy due to rotation of molecule i.e. Rotational energy, E_r.

    (iii) Energy due to vibration of molecule i.e. Vibrational energy, E_v.

    $\therefore$ Total energy, $E = E_e + E_r + E_v$

    The magnitudes of these energies are sufficiently different and

    $$E_e > E_v > E_r$$

    Hence, we can classify molecular spectra into **three distinct regions.**

1. Pure Rotational Spectrum :

The transitions between the rotational energy states of a molecule with the emission or absorption of a photon give rise to pure rotational spectrum.

In this case, electronic configuration of the atom and vibrational state of molecule remains the same.

Rotational levels are separated by small energy interval of the order of 10^{-3} eV. Pure rotational spectra lie in microwave or far infrared region.

2. Rotational - Vibrational Spectrum :

The transitions between vibrational energy states of a molecule with the emission or absorption of a photon, when electronic configuration of an atom remains the same, giving rise to rotational-vibrational spectrum.

If the rotational state of molecule do not change, we get a pure vibrational spectrum. Vibrational spectra lie in near IR region. Vibrational states are separated by energy interval of about 0.1 eV.

3. Electronic Spectrum :

The transitions between two electronic energy states of an atom in a molecule with emission or absorption of photon give rise to electronic spectrum. The energy separation between different electronic states is in between 1 eV - 10 eV. The electronic spectra lie in visible and ultravisible part of spectrum.

6.1 Rotational Energy Level   (April 17)

- The rotational motion naturally will involve the moment of inertia of a molecule. Hence the study of the rotational spectra can be used for the analysis in many scientific and industrial applications. We can use the rotational spectra to distinguish the presence of isotope in a sample. It can even detect different isomers, provided, they have different moment of inertia. Rotational spectra can also be used in the chemical examination of interstellar space. It has been used to determine some stable molecules in the space like water, ammonia, etc. which helps in predicting the presence or origin of the biological molecule. By comparing the intensities of the rotational transitions, temperature of the interstellar material can be estimated.

- Like any energy levels of electrons in atoms, the rotational energy levels are quantized. One can mathematically calculate the energy of the levels due to the rotation of the molecule. The system becomes more complicated with more number of molecules and the type of bonding between the atoms. Hence for simplicity, we consider a diatomic molecule as a rigid rotator.

- Only the molecules, having permanent electric dipole moment, can give rise to pure rotational spectra. Hence molecules like HCl, HBr, HI, H_2O, etc. show the rotational spectra in far infrared region. Homonuclear molecules such as H_2, O_2, N_2 also show rotational Raman spectra because of the polarizability of the molecule.

The Rigid Diatomic Molecule :   **(Oct. 15, April 17, 11)**

- Consider the simplest linear diatomic molecule as shown in Fig. 6.1. Let m_1 and m_2 be the mass of each atom. Let the molecules are joined by a bond such that the distance between them remains constant. The bond length is given by

$$r_0 = r_1 + r_2 \qquad \qquad \text{... (6.1)}$$

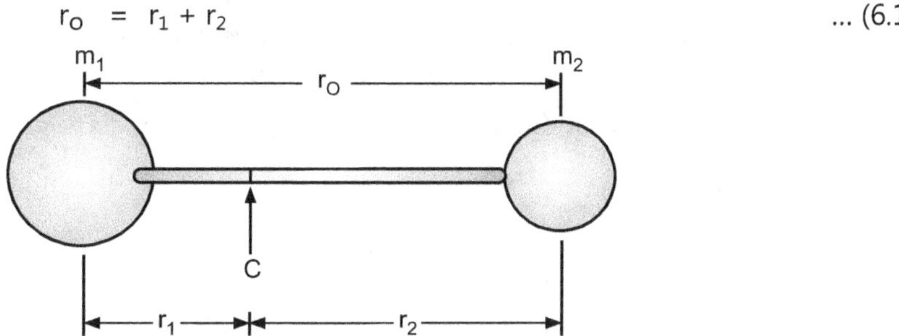

Fig. 6.1 : A rigid diatomic molecule treated as two masses,

m_1 and m_2 joined by a rigid bar of length $r_0 = r_1 + r_2$

- The molecule rotates around an axis passing through centre of gravity C or centre of mass of the system and perpendicular to the line joining m_1 and m_2 respectively from centre of mass, then

$$m_1 \, r_1 \; = \; m_2 \, r_2 \qquad \text{... (6.2)}$$

The moment of inertia of the diatomic molecule about a point C is given by

$$I \; = \; m_1 \, r_1^2 + m_2 \, r_2^2$$

By using equation (6.2) in the above equation, we have

$$I \; = \; m_2 \, r_2 \, r_1 + m_1 \, r_1 \, r_2$$

$$= \; r_1 \, r_2 \, (m_1 + m_2) \qquad \text{... (6.3)}$$

From equations (6.1) and (6.2), we have

$$m_1 \, r_1 \; = \; m_2 \, r_2$$

$$= \; m_2 \, (r_0 - r_1)$$

$$\therefore \qquad (m_1 + m_2) \, r_1 \; = \; m_2 \, r_0$$

$$\therefore \qquad r_1 \; = \; \frac{m_2}{m_1 + m_2} \, r_0 \qquad \text{... (6.4)}$$

Similarly, $\qquad r_2 \; = \; \dfrac{m_1}{m_1 + m_2} \, r_0 \qquad \text{... (6.5)}$

Substituting equations (6.4) and (6.5) in equation (6.3), we get

$$I \; = \; \frac{m_1 \, m_2}{m_1 + m_2} \, r_0^2$$

$$= \; \mu \, r_0^2 \qquad \text{... (6.6)}$$

where $\mu = \dfrac{m_1 \, m_2}{m_1 + m_2}$ is known as *reduced mass of the system*. Here we have converted two-body problem into a single body. Hence we can say that the rotation of a diatomic molecule is equivalent to the rotation of a single particle.

- If the molecule is rotating with angular velocity ω, then classically it has energy equal to $\frac{1}{2} I \omega^2$. The angular momentum $L = I \omega$.

$$\therefore \qquad E \; = \; \frac{L^2}{2I} \qquad \text{... (6.7)}$$

According to quantum mechanics, the angular momentum is quantized and given by

$$L = \sqrt{J \, (J + 1)} \; \hbar \quad \text{where } \hbar = \frac{h}{2\pi}$$

Equation (6.7) becomes

$$E_J = \frac{h^2}{8\pi^2 I} J (J + 1) \text{ joules} \qquad \text{where } J = 0, 1, 2, \ldots \qquad \ldots (6.8)$$

$\therefore$ Rotational energy level,

$$E_J \approx \frac{J (J + 1)\, \hbar^2}{2I}$$

where h is Planck's constant and J is known as *rotational quantum number*. J takes only integral values from zero upwards.

Equations (6.7) and (6.8) show that

(i)   Diatomic molecule cannot have any arbitrary value of rotational energy. The energy is quantized and is limited by rotational quantum number J.

(ii) The rotational energy depends upon moment of inertia i.e. on mass, shape and size of molecule.

Rotation about Bond Axis :

*   In previous case we have considered only rotation about an axis perpendicular to the bond axis of diatomic molecule. What about rotations about axis of symmetry itself ?

*   Such rotations are neglected because mass of atom is located almost entirely in its nucleus whose radius is only 10^{-4} of the radius of an atom itself. The main contribution to I of diatomic molecule about bond axis comes from electron which are concentrated in a region whose radius is roughly half the bond length R, but whose mass is $\frac{1}{4000}^{\text{th}}$ of total molecular mass. As allowed rotational energy levels are proportional to $\frac{1}{I}$, hence the rotation about the symmetry axis must involve energies $\approx 10^4$ times E_J values than previous one. Hence energies of several eV would be involved in any rotation about the symmetry axis of diatomic molecule. Bond energy of this order make the molecule to dissociate in any environment in which such rotations are excited.

Rotational Spectra of Rigid Diatomic Molecule :   **(April 16, 11)**

*   When we deal with the spectra due to the transitions from one energy level to other energy level, the differences between these energies are important. If ΔE is the difference between the energy levels, then the corresponding frequency of radiation emitted or absorbed is given by

$$\upsilon = \frac{\Delta E}{h} \text{ in Hz}$$

*   It can be expressed in terms of wave number as

$$\bar{\upsilon} = \frac{\Delta E}{ch} \text{ cm}^{-1} \qquad\qquad \because \bar{\upsilon} = \frac{1}{\lambda} = \frac{\upsilon}{c}$$

where c is the velocity of light.

- In general, the rotational spectra are discussed in terms of wave number, hence let us consider energies expressed in these units. By using equation (6.8), we get

$$\varepsilon_J \;=\; \frac{E_J}{ch} \;=\; \frac{h}{8\pi^2\,Ic}\, J\,(J+1) \text{ in cm}^{-1} \qquad \text{... (6.9)}$$

$$\therefore \qquad \varepsilon_J \;=\; BJ\,(J+1)\text{ cm}^{-1} \text{ where } J = 0, 1, 2, \ldots \qquad \text{... (6.10)}$$

where B is known as *rotational constant* which is given $(\bar{\upsilon})$ as

$$B \;=\; \frac{h}{8\pi^2\,Ic}\text{ cm}^{-1} \qquad \text{... (6.11)}$$

- By using equation (6.11) we can show the energy levels diagrammatically as shown in Fig. 6.2. For J = 0, ε_J = 0, hence the molecule is not rotating at all. For J = 1, the rotational energy level is 2B and rotating molecule has its lowest angular momentum. In this manner we may continue to calculate ε_J with increasing J values. With the increase in the value of J, value of ε_J increases. At normal temperatures, the centrifugal force of rapidly rotating molecule is always less than the strength of the bond, hence the molecular bond is stable.

- Now, consider difference between the levels to discuss the spectrum. Imagine that the molecule is raised from energy level J = 0 to J = 1 state by absorption of energy.

Energy absorbed will be

$$\varepsilon_{J=1} - \varepsilon_{J=0} \;=\; 2B - 0 = 2B \text{ cm}^{-1}$$

$$\therefore \qquad \bar{\upsilon}_{J=0 \rightarrow J=1} \;=\; 2B \text{ cm}^{-1}$$

The absorption line will appear at 2B cm^{-1}.

If the molecule is raised from J = 1 to J = 2 level by absorption of more energies, then

$$\bar{\upsilon}_{J=1 \rightarrow J=2} \;=\; \varepsilon_{J=2} - \varepsilon_{J=1} = 6B - 2B$$

$$= 4B \text{ cm}^{-1}$$

In general,
$$\bar{\upsilon}_{J \rightarrow J+1} \;=\; \varepsilon_{J+1} - \varepsilon_J$$

$$= B\,(J+1)\,(J+2) - BJ\,(J+1)$$

$$= B\,[J^2 + 3J + 2 - J^2 - J]$$

$$= B\,(2J+2)$$

$$\therefore \qquad \bar{\upsilon}_{J \rightarrow J+1} \;=\; 2B\,(J+1)\text{ cm}^{-1} \qquad \text{... (6.12)}$$

- The energy level transitions result in an absorption spectra having lines at 2B, 4B, 6B, 8B, ... cm^{-1}. We can also get an identical emission spectra having lines at 2B, 4B, 6B, 8B, ... cm^{-1}. It is shown at the bottom of Fig. 6.3. It is clear that in the case of rotational spectra, we get the lines which are equally spaced from each other.

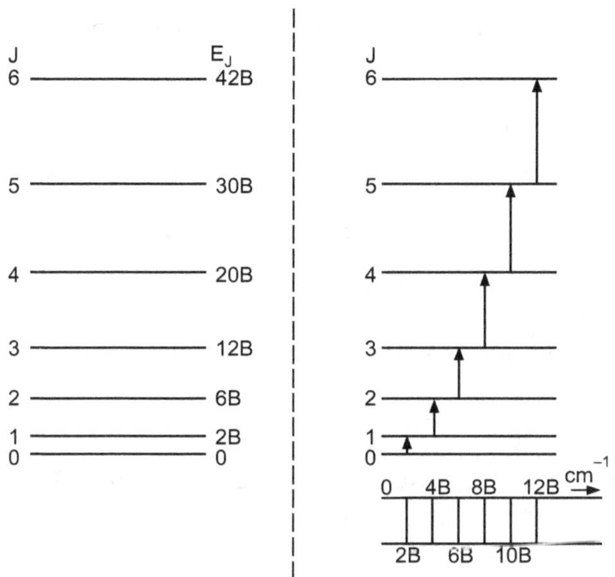

Fig. 6.2 : Allowed rotational energy levels of a rigid diatomic molecule

Fig. 6.3 : Allowed transitions between the energy levels of a rigid diatomic molecule

- The Schrödinger's wave equation shows that for rigid molecule, transitions are such that J changes by unity only, other transitions are forbidden. Hence the selection rules for the rigid diatomic rotator are

$$\Delta J = \pm 1$$

- This type of spectrum will be observed only in the case of heteronuclear molecule i.e. HCl, CO, etc. as they have changing dipole moment. If molecule is homonuclear, there will be no dipole component change during the rotation, hence no interaction with radiation and molecule like N_2 and O_2 will not show a rotational spectrum.

6.2 Vibrational Energy Levels   (Oct. 17, 16)

- Let us consider the vibrational motion of a diatomic molecule. Suppose two atoms forming the molecule have masses m_1 and m_2 and are separated by an equilibrium distance R_0 between them. As the molecule vibrates along the line joining the atoms, the distance between them changes.

- Let at any instant during the vibration the separation between m_1 and m_2 be R. Therefore, there is change in the distance between two atoms is equal to $(R - R_0)$. The restoring force acting on atoms is directly proportional to the deviation from the equilibrium position R_0.

Hence, $F \propto - (R - R_0)$ ... (6.13)

- The negative sign indicates that the force is directed towards the centre of mass. Let $R - R_0 = x$, then equation (6.13) gives

$$F \propto -x$$

or $$F = -kx \qquad \qquad \ldots (6.14)$$

where k is force constant and is a measure of the stiffness of bond.

But $$F = ma = m\frac{d^2x}{dt^2} \qquad \qquad \ldots (6.15)$$

From equations (6.14) and (6.15), we get

$$m\frac{d^2x}{dt^2} = -kx$$

or $$\frac{d^2x}{dt^2} = -\frac{k}{m}x \qquad \qquad \ldots (6.16)$$

- In diatomic molecule we have two atoms of masses m_1 and m_2 joined by a bond which may be compared to an elastic spring as shown in Fig. 6.4 (a).

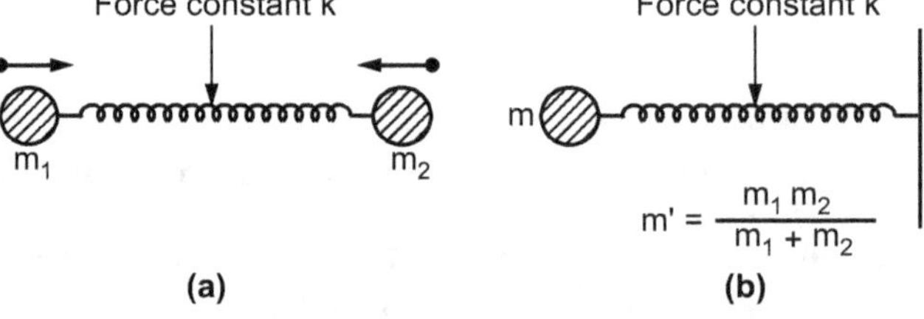

Fig. 6.4 : Two-body oscillator

- As there is no external force acting on the system, the linear momentum remains constant. The oscillations of two atoms, therefore, do not change the position of centre of mass i.e. m_1 and m_2 vibrate back and forth, relative to centre of mass in opposite directions and both reach the extremes of their respective motion at the same time. Such a two body oscillator is equivalent to motion of a single harmonic oscillator of mass equal to the reduced mass of the two body system.

- Therefore m in equation (6.16) should be replaced by m' which is reduced mass, given by

$$m' = \frac{m_1 m_2}{m_1 + m_2}$$

as shown in Fig. 6.4 (b).

- Now substituting $m = m'$ in equation (6.16), we have

$$\frac{d^2x}{dt^2} = -\frac{k}{m'}x \qquad \qquad \ldots (6.17)$$

- This is the equation of simple harmonic motion and according to classical physics, the frequency of oscillation is given by

$$\upsilon_0 = \frac{1}{2\pi}\sqrt{\frac{k}{m'}}$$

- The potential energy of the system is related to the force by the relation

$$F = -\frac{dV}{dx} \qquad \text{... (6.18)}$$

But according to equation (6.13),

$$F = -kx$$

$$\therefore \qquad \frac{dV}{dx} = kx$$

or

$$dV = kx\,dx \qquad \text{... (6.19)}$$

Hence, potential energy for a displacement x from the equilibrium position $x = 0$ is given by

$$V = \int_0^x kx\,dx$$

or

$$V = \frac{1}{2}kx^2 \qquad \text{... (6.20)}$$

- In order to obtain energy levels of simple harmonic oscillator, consider Schrodinger's one dimensional, time independent equation,

$$\frac{d^2\psi}{dx^2} + \frac{2m}{\hbar^2}(E - V)\,\psi = 0 \qquad \text{... (6.21)}$$

- Substituting $m = m'$ and $V = \frac{1}{2}kx^2$ for the molecular oscillator, Schrodinger's equation becomes

$$\frac{d^2\psi}{dx^2} + \frac{2m'}{\hbar^2}\left(E - \frac{1}{2}kx^2\right)\psi = 0$$

or

$$-\frac{\hbar^2}{2m'}\frac{d^2\psi}{dx^2} + \frac{1}{2}kx^2\psi = E\psi \qquad \text{... (6.22)}$$

- The energy levels of simple harmonic oscillator are now given by the solution of equation (6.22), which are

$$E_\upsilon = \left(\upsilon + \frac{1}{2}\right)h\upsilon_0 \qquad \text{... (6.23)}$$

where

$$\upsilon_0 = \frac{1}{2\pi}\sqrt{\frac{k}{m'}}$$

and $\upsilon = 0, 1, 2, \ldots$ etc. is vibrational quantum number.

For ground state $\upsilon = 0$ and hence ground state vibrational energy

$$E_0 = \frac{1}{2}h\upsilon_0$$

where υ_0 is the classical frequency.

Vibrational energy levels are equally spaced :

- The energy level given by equation (6.23) are called vibrational energy levels of the molecule. Vibrational energy levels corresponding to vibrational quantum number $\upsilon = 0, 1, 2, 3, \ldots$ etc. are shown in Fig. 6.5.

$$\upsilon = 3 \underline{\hspace{3cm}} \quad E_3 = \frac{7}{2} h\upsilon_o$$

$$\upsilon = 2 \underline{\hspace{3cm}} \quad E_2 = \frac{5}{2} h\upsilon_o$$

$$\upsilon = 1 \underline{\hspace{3cm}} \quad E_1 = \frac{3}{2} h\upsilon_o$$

$$\upsilon = 0 \underline{\hspace{3cm}} \quad E_0 = \frac{1}{2} h\upsilon_o$$

Fig. 6.5 : Vibrational energy levels

From equation (6.23),

For $\upsilon = 0$

$$E_0 = \frac{1}{2} h\upsilon_0$$

For $\upsilon = 1$ $\qquad E_1 = \frac{3}{2} h\upsilon_0$

For $\upsilon = 2$ $\qquad E_2 = \frac{5}{2} h\upsilon_0 \ \ldots \ \text{etc.}$

- The energy levels are equally spaced. The energy of the ground state is not zero as in case of rotational energy levels but has a value $\frac{1}{2} h\upsilon_0$.

- This is in accordance with uncertainty principle, because if the molecule has zero vibrational energy i.e. it is at rest, the uncertainty in position is $\Delta x = 0$.

As $\qquad \Delta x \, \Delta P_x = \hbar$

$\therefore \qquad \Delta P_x = \dfrac{\hbar}{\Delta x}$

$\therefore \qquad \Delta P_x = \dfrac{\hbar}{0} = \infty$

- Hence uncertainty in momentum is $\Delta P_x = \infty$. But a particle with $E = 0$ cannot have infinitely uncertain momentum.

- As vibrational energy levels belong to energy values,

$$E_0 = \frac{1}{2} h\upsilon_0$$

$$E_1 = \frac{3}{2} h\upsilon_0 \ \ldots \ \text{etc.}$$

Therefore, $\qquad E_1 - E_0 = h\upsilon_0$

Similarly, $\qquad E_2 - E_1 = h\upsilon_0$

In general, $\qquad E_1 - E_0 = E_2 - E_1 = E_3 - E_2 = \ldots = h\upsilon_0$

- Thus vibrational energy levels are equally spaced with energy interval $h\upsilon_0$.

Selection Rules : $\hfill$ (Oct. 15)

- The selection rule for transitions between vibrational energy levels is "only those transitions are allowed between vibrational energy levels for which the vibrational quantum number υ changes by one unit".

  i.e. $\qquad\qquad \Delta\upsilon = \pm 1$

- The oscillating dipole, whose frequency is υ_0 can only absorb or emit electromagnetic radiations of the same frequency. All quanta of frequency υ_0, have the energy $h\upsilon_0$. Thus oscillating dipole can only absorb an energy $\Delta E = h\upsilon_0$ at one time, so that its energy increases from $\left(\upsilon + \dfrac{1}{2}\right) h\upsilon_0$ to $\left(\upsilon + \dfrac{1}{2} + 1\right) h\upsilon_0$. Similarly, it can only emit an energy $\Delta E = h\upsilon_0$.

  Hence selection rule is

  $$\Delta\upsilon = \pm 1$$

Frequency of lines in vibrational spectra :

If the transition occurs from an energy level with quantum number $(\upsilon + 1)$ to an energy level with quantum number υ, then

$$h\upsilon = \Delta E = E_{\upsilon + 1} - E_{\upsilon}$$

$$= \left(\upsilon + 1 + \frac{1}{2}\right) h\upsilon_0 - \left(\upsilon + \frac{1}{2}\right) h\upsilon_0$$

$$h\upsilon = h\upsilon_0$$

or $\qquad\qquad\qquad \upsilon = \upsilon_0$

But $\qquad\qquad\qquad \upsilon_0 = \dfrac{1}{2\pi} \sqrt{\dfrac{k}{m'}}$

$\therefore \qquad\qquad\qquad \upsilon = \upsilon_0 = \dfrac{1}{2\pi} \sqrt{\dfrac{k}{m'}}$ $\hfill$ ... (6.24)

- This gives the frequency of spectral line in vibrational emission spectra. Similarly, the frequency of spectral line in vibrational absorption spectra is also υ_0. Thus the frequency of radiation emitted or absorbed is equal to the classical frequency υ_0 irrespective of the values of vibrational quantum number.

- The frequency of lines in vibrational spectra lies in the infra-red region with wavelength ranging from 8000 A° to 50,000 A° for a number of molecules.

e.g. For a vibrating HCl molecule the force constant $k = 470 \ Nm^{-1}$.

The reduced mass
$$m' = \frac{m_1 m_2}{m_1 + m_2}$$

$$= \frac{1 \times 35}{1 + 35} \times 1.67 \times 10^{-27}$$

or
$$m' = 1.62 \times 10^{-27} \ kg$$

Now, frequency,
$$\upsilon = \frac{1}{2\pi} \sqrt{\frac{k}{m'}}$$

$$= \frac{1}{2 \times 3.142} \sqrt{\frac{470}{1.62 \times 10^{-27}}}$$

$$\therefore \quad \upsilon = 8.6 \times 10^{13} \ Hz$$

Hence, wavelength,
$$\lambda = \frac{c}{\upsilon} = \frac{3 \times 10^8}{8.6 \times 10^{13}}$$

$$\lambda = 35000 \ A°$$

This wavelength lies in the infra-red region.

Potential energy versus inter-atomic distance curve :

- The vibrational rotational energy levels of a diatomic molecule, on a potential energy versus inter-atomic distance curve, are shown in Fig. 6.6.

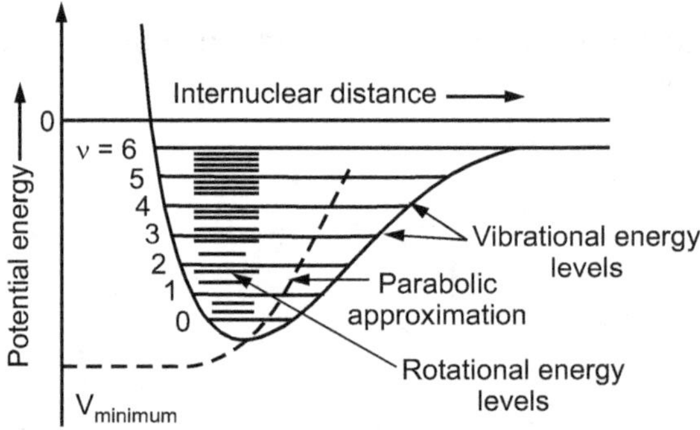

Fig. 6.6 : Potential energy of a diatomic molecule as a function of interatomic distance

- As vibrational quantum number υ increases, the higher vibrational states of the molecule do not obey the relation

$$E = \left(\upsilon + \frac{1}{2}\right) h\upsilon_0$$

$$\upsilon_0 = \frac{1}{2\pi} \sqrt{\frac{k}{m'}}$$

where $\quad k -$ force constant of the molecule

and $\quad m' -$ reduced mass.

- If we look at the potential energy V versus inter-nuclear distance R curve, we find that in the neighbourhood of the minimum of the curve, which corresponds to the normal configuration of the molecule, the shape of the curve is very nearly a parabola.

- Therefore, for lower vibrational states of the molecule, the potential energy curve is parabolic but for higher vibrational states i.e. with increasing energy, the potential energy curve deviates more and more from parabolic form, so that the energy well is wider than the parabola. As a result, the spacing between adjacent energy levels for higher vibrational quantum number is less than the spacing between adjacent energy levels for lower values of vibrational quantum number, so that for higher quantum numbers, the energy levels are closer.

6.3 Vibration - Rotation Spectra (April 12)

- The pure vibrational spectra are observed only in liquid where interaction between adjacent molecules inhibit rotation. As the excitation energies involved in molecular rotation are much smaller than those involved in vibration, freely moving molecules in gas or vapour always rotate irrespective of their vibrational states. Spectra obtained from such molecule consist of closely spaced lines due to transitions between various rotational states of one vibrational level and rotational states of other. The lines in such spectra are appeared as a broad streak, called a vibration - rotation band.

- Spectra in the operation of the tunable dye layer lies in this category. Such dye layer uses an organic dye whose molecules are pumped to excited states by another laser. The dye gives fluorescence in broad emission spectra. From this band light of desired λ can be selected for light amplification with the help of two facing mirrors (one of them is partly transparent). The distance between mirrors is kept to be in integral multiple of $\lambda/2$. Trapped laser light forms optical standing wave which emerges through partly transparent mirror.

6.4 Electronic Spectra of Molecules

- In rotational or vibrational spectra, the energies are due to motion of atomic nuclei. The electrons in molecule can be excited to higher energy levels than those corresponding to its ground state. The spacing between two levels is much greater than spacing in rotational or vibrational energy levels.

- Electronic transitions involve radiation in visible and ultraviolet region. Each transition appears as a series of closely spaced lines called a band due to presence of different rotational and vibrational states in each electronic state. All molecules exhibit electronic spectra as a dipole moment change always accompanies a change in electronic configuration of a molecule. Hence homonuclear molecules such as H_2, N_2 which have neither rotational nor vibrational spectra have electronic spectra.

Fluorescence   (April 16, Oct. 15)

A molecule in an excited state can lose energy and return to ground states in various ways. These are

1. Molecule may emit a photon of same frequency as that of the photon it absorbed and come to ground state in single step.

2. **Fluorescence :** Here molecule gives up some of its vibrational energy in collision with other molecules so that the downward relative transition originates from a lower vibrational level in upper electronic state. (Refer Fig. 6.7) Hence fluorescent radiation has lower frequency than that of absorbed radiation.

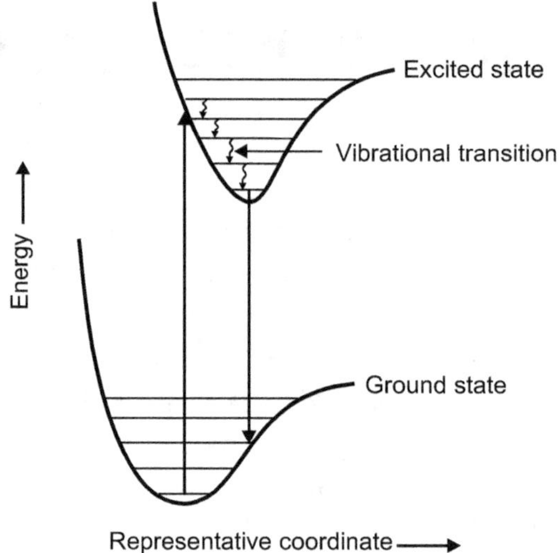

Fig. 6.7 : Origin of fluorescence

Fluorescence has many applications :

1. It is used to identify minerals and biochemical compound.

2. Fabric brightners that are added to detergents absorb U.V. radiation and then fluoresce blue light.

3. In a fluorescent lamp, a mixture of mercury vapour and inert gas such as argon inside a glass tube gives off U.V. radiation when an electric current is passed through it.

Inside of tube is coated with fluorescent material (called phosphor). It gives out visible light when excited through U.V. radiation.

Phosphorescence

•   As shown in Fig. 6.8 the moelcule in its singlet (total spin quantum number S = 0) ground state absorbs photon and is raised to singlet excited state.

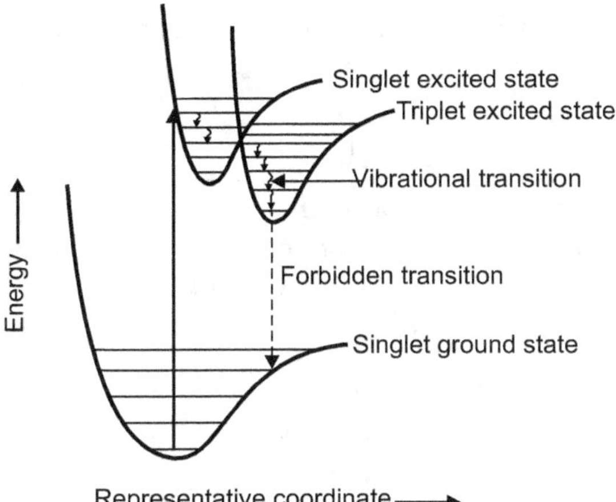

Fig. 6.8 : The origin of phosphorescence

- In collision the molecule can undergo radiationless transitions to lower vibrational level that may happen to have about same energy as one of the levels in triplet (S = 1) excited state. There is probability for shift to triplet state to occur. Collision in triplet state brings molecular energy below that of cross-over point, so that it is trapped in triplet states and ultimately reaches V = 0 level.

- Transition from triplet to singlet state is not allowed by selection rule. (Forbidden transition). Such transitions are long life lived. Hence phosphorescence radiation may be emitted minute or hours after the initial absorption.

Solved Problems

Problem 6.1 : *The force constant of the bond in CO molecule is 1956 N/m. Calculate the frequency of vibration of the molecule and the spacing between its vibrational energy levels in eV. Given : Reduced mass of CO is 1.16 × 10^{-26} kg, h = 6.63 × 10^{-27} erg-sec. and 1 eV = 1.6 × 10^{-12} erg., c = 3 × 10^8 m/s.* **(April 17)**

Solution : The frequency of vibration is given by,

$$\upsilon_0 = \frac{1}{2\pi}\sqrt{\frac{k}{\mu}}$$

$$= \frac{1}{2\pi}\sqrt{\frac{1956 \text{ N/m}}{1.16 \times 10^{-26} \text{ kg}}}$$

$$= \mathbf{6.535 \times 10^{13} \, Hz} \qquad\qquad \textbf{... Ans.}$$

The separation between two successive vibrational energy levels is

$$\Delta\varepsilon = \varepsilon_{V+1} - \varepsilon_V = h\upsilon_0$$

$$= 6.63 \times 10^{-27} \times 6.535 \times 10^{13}$$

$$= \mathbf{4.333 \times 10^{-13} \, erg.} \qquad\qquad \textbf{... Ans.}$$

The spacing between vibrational energy levels in eV is given by

$$\Delta E = \frac{4.333 \times 10^{-13}}{1.6 \times 10^{-12}}$$

$$= 0.271 \text{ eV} \qquad \text{... Ans.}$$

Problem 6.2 : *In the near infrared spectrum of HCl molecule, a single band at 2856 cm^{-1} is observed, due to the transition between the vibrational levels. Show that the force constant K is 4.79 $\times 10^5$ dyne/cm if mass of hydrogen is 1.7 $\times 10^{-24}$ gm.*

Solution : The force constant is given by,

$$K = 4\pi^2 \mu\, c^2\, \omega_e^2$$

Here,

$$\omega_e = 2856 \text{ cm}^{-1}$$

$$\mu = \frac{M_H\, M_{Cl}}{M_H + M_{Cl}}$$

$$= \frac{1 \times 35}{1 + 35}\, M_H$$

$$= \frac{35}{36} \times 1.7 \times 10^{-24}$$

$$= 1.653 \times 10^{-24} \text{ gm}$$

$$\therefore \qquad K = 4\pi^2 \times 1.653 \times 10^{-24} \times (3 \times 10^{10})^2 \times 2856^2$$

$$= \mathbf{4.791 \times 10^5 \text{ dyne/cm}} \qquad \text{... Ans.}$$

Problem 6.3 : *The mean of the intermolecular distances for HCl^{35} in the v = 0 and v = 1 levels is 1.293 A°. Calculate the difference in cm^{-1} between P and R branches of the fundamental band for HCl^{35}.*

Given : *h = 6.62 $\times 10^{-27}$ erg- sec., c = 3 $\times 10^{10}$ cm/s.*

Solution : The reduced mass of HCl molecule is

$$\mu = \frac{(1 \times 35)/(6.023 \times 10^{23})^2}{(1 + 35)/(6.023 \times 10^{23})}$$

$$= \frac{35}{36 \times (6.023 \times 10^{23})}$$

$$= 1.614 \times 10^{-24} \text{ gm}$$

The rotational constant for the molecule is

$$B = \frac{h}{8\pi^2 I c}$$

$$= \frac{h}{8\pi^2\, \mu r^2 c}$$

$$= \frac{6.62 \times 10^{-27} \text{ erg-sec.}}{8 \times \pi^2 \times (1.61 \times 10^{-24}) \times (1.293 \times 10^{-8} \text{ cm})^2 \times (3 \times 10^{10})}$$

$$= 10.383 \text{ cm}^{-1}$$

We know that when the interactions between vibration and rotation are ignored, then the line in P, Q, R branches in terms of wave number is given by,

$$\bar{v}_{spect} = \bar{v}_0 + 2Bm$$

For P branch, m = − 1 and for R branch, m = + 1.

$$\therefore \qquad \bar{v}_{P\text{-branch}} = \bar{v}_0 - 2B$$

$$\bar{v}_{R\text{-branch}} = \bar{v}_0 + 2B$$

Hence, the difference in the P and R branches of the fundamental band is given by,

$$\bar{v}_{R\text{-branch}} - \bar{v}_{P\text{-branch}} = 4B$$

$$= 4 \times 10.383$$

$$= \textbf{41.532 cm}^{-1} \qquad\qquad \textbf{... Ans.}$$

Problem 6.4 : *When CO is dissolved in liquid carbon tetrachloride, IR radiation of frequency* 6×10^{13} *Hz is absorbed. Carbon tetrachloride is transparent itself at this frequency so that absorption must be due to CO.*

(a) *What is the force constant of the bond in the CO molecule ? Given reduced mass of carbon is* 1.14×10^{-26} *kg.*

(b) *What is spacing between its vibrational energy levels ?*

Solution : (a)

$$v_o = \frac{1}{2\pi}\sqrt{\frac{K}{\mu}}$$

$$K = 4\pi^2\, v_o^2 \times \mu$$

$$= 4\pi^2 \times (6 \times 10^{13})^2 \times (1.14 \times 10^{-26})$$

$$= \textbf{1.62} \times \textbf{10}^3 \textbf{ N/m} \qquad\qquad \textbf{... Ans.}$$

(b) Separation ΔE between vibrational levels in CO is

$$\Delta E = E_{v+1} - E_v = hv_o$$

$$= 6.63 \times 10^{-34} \text{ Js} \times 6 \times 10^{13} \text{ Hz}$$

$$= 3.978 \times 10^{-20} \text{ J}$$

$$= \textbf{0.2486 eV} \qquad\qquad \textbf{... Ans.}$$

Problem 6.5 : *The spacing between vibrational level of CO molecule is 0.08 eV. Calculate the value of force constant. Take mass of carbon atom = 12 and that of oxygen 16 times mass of proton =* 1.67×10^{-27} *kg.*   **(April 12)**

Solution : Reduced mass of CO molecule,

$$\mu = \frac{m_1\, m_2}{m_1 + m_2}$$

$$= \frac{12 \times 16}{12 + 16} \times 1.67 \times 10^{-27}$$

$$= 1.14 \times 10^{-26} \text{ kg}$$

$$\Delta E = E_{\upsilon+1} - E_\upsilon = h\upsilon_o = \frac{h}{2\pi}\sqrt{\frac{K}{\mu}} = 0.08 \text{ eV} = 0.08 \times 1.6 \times 10^{-19} \text{ J}$$

$$\frac{K}{\mu'} = \left(\frac{0.08 \times 1.6 \times 10^{-19} \times 2\pi}{h}\right)^2$$

$$K = \left(\frac{0.08 \times 1.6 \times 10^{-19} \times 2\pi}{6.63 \times 10^{-34}}\right)^2 \times 1.14 \times 10^{-26}$$

$$= \textbf{167.8 Nm}^{-1} \qquad \qquad \text{... Ans.}$$

Summary

1. Molecular spectroscopy is the study of interaction of e.m. wave and matter which gives more information regarding structure of matter.

2. There are three types of molecular spectra :

    Pure rotational, Rotational - Vibrational and Electronic spectra.

3. In rotational spectra, rotational energy level is given as

    $$E_J = \frac{J\,(J + 1)\,\hbar^2}{2I}$$

4. Rotational energy in terms of wave number

    $$\varepsilon_J = \frac{E_J}{ch} = BJ\,(J + 1) \text{ cm}^{-1}, \text{ where } J = 0, 1, 2, \ldots\ldots$$

    B is known as rotational constant.

    $$B = \frac{h}{8\pi^2\, Ic} \text{ cm}^{-1}$$

5. Vibrational energy level

    $$E_\upsilon = \left(\upsilon + \frac{1}{2}\right) h\upsilon_o$$

    where υ = Vibrational quantum number

    $$= 0, 1, 2, \ldots\ldots$$

    and $\Delta\upsilon = \pm\, 1$

6. In case of diatomic molecule,

$$E_v = \left(v + \frac{1}{2}\right) \hbar \sqrt{\frac{K}{\mu}}$$

7. Fluorescence and phosphorescence are due to electronic spectra.

Exercise

(A) Short Answer Type Questions :

1. What are three major types of molecular spectra ?
2. Define fluorescence and phosphorescence.
3. Define reduced mass of a molecule.
4. What is vibrational quantum number ?
5. What is vibrational - rotational spectra ?

(B) Long Answer Type Questions :

1. Obtain an expresssion for rotational energy level of rigid diatomic molecule.
2. Show that for rigid diatomic molecule

$$E_J = \frac{J(J+1)\,\hbar^2}{2I}$$

3. Show that vibrational energy level is given by

$$E_v = \left(v + \frac{1}{2}\right) h\nu_o$$

$$= \left(v + \frac{1}{2}\right) \hbar \sqrt{\frac{K}{\mu}}$$

4. Discuss vibrational-rotational spectra in detail.
5. What is electronic spectra of molecule ? Hence discuss fluorescence and phosphorescence in detail.

(C) Numerical Problems :

1. Carbon monoxide (CO) molecule has a bond length R of 0.113 nm and the masses of the ^{12}C and ^{16}O atoms are respectively 1.99×10^{-26} kg and 2.66×10^{-26} kg. Find

   (a) energy and

   (b) angular velocity of CO molecule when it is in its lowest rotational state.

   Hint : $I = \mu R^2$ and $E_J = \dfrac{J(J+1)\,\hbar^2}{2I}$ (**Ans.** (a) 4.76×10^{-4} eV, (b) $\omega = 3.23 \times 10^{11}$ rad/s)

2. In CO, the $J = 0 \rightarrow J = 1$ absorption line occurs at frequency 1.15×10^{11} Hz. What is the bond length of CO molecule ?   (**Ans.** 0.113 m)

3. The lowest vibrational states of ^{23}Na ^{35}Cl molecule are 0.063 eV apart. Find the approximate force constant of the molecule. (**Ans.** 2.1×10^2 N/m)

4. The force constant of ^{1}H ^{19}F molecule is approximately 966 N/m. Find frequency of vibration of molecule. (**Ans.** 1.24×10^{14} Hz)

5. The force constant of CO bond is 187 N/m. Find the frequency of vibration of CO molecule and spacing between vibrational levels.

    (Mass of ^{12}C atom = 1.99×10^{-26} kg and ^{16}O = 2.66×10^{-26} kg.)

    (**Ans.** $\upsilon = 2.04 \times 10^{13}$ Hz, $\Delta E = 8.44 \times 10^{-2}$ eV)

Chapter **7**...

Raman Spectroscopy

Contents ...

Dr. C.V. Raman

In 1928, Sir **Dr. C.V. Raman** while studying scattering of light found that when a beam of monochromatic light was passed through organic liquid such as benzene, toluene, etc. the scattered light contained other frequencies in addition to that incident light. Such effect had been predicted theoretically by Adolf Smekal in 1923. In 1922, Sir Dr. C.V. Raman published his work on the "Molecular Diffraction of Light," the first of a series of investigations at that time which led to the discovery of Raman effect (on 28 February 1928). Later, he was awarded Nobel Prize in 1930 for the discovery of Raman effect.

Introduction

• In this chapter, we shall study classical and quantum theory of Raman effect, experimental set up of it and application of Raman spectroscopy.

RAMAN EFFECT (April 16; Oct. 16, 15)

• Rayleigh scattering of light is a familiar phenomenon which is responsible for the blue color of the sky. Light from the sun strikes air molecules and is scattered in all directions without losing energy or changing frequency. The blue color arises because the strength of the scattering depends strongly on the fourth power of the frequency. Thus, blue light is much more strongly scattered than yellow or red.

- In 1928, C.V. Raman discovered another type of light scattering by particles in which the frequency changes when the light is scattered. When monochromatic radiation passes through pure transparent substance (i.e. solid, liquid or gas), it may undergo scattering in addition to absorption and transmission.

- The scattered radiations consist of :

  1. Central line called as Rayleigh line, frequency of such line is same as frequency of source (υ_0).

  2. A series of displaced lines with slightly different frequencies on both sides of central line. This phenomenon of scattering with modified frequency is called as **Raman scattering.**

- Raman scattering is weak scattering and is about 1 % of the incident radiation or less than 0.01 per cent of Rayleigh scattering. Raman scattering does not involve absorption of radiation. The scattering occurs in 10^{-12} sec after excitation. Raman spectroscopy deals with the measurement of intensity of scattered radiation as a function of wavelength.

- The lines in Raman scattering are frequently arranged symmetrically on both sides of the Rayleigh line. The frequency shift $\Delta\upsilon$ occurs when some of the energy of the scattered photon is taken up by a molecule, which is excited into vibrational motion. The lines of frequency less than that of Rayleigh line ($\upsilon < \upsilon_0$) are called as *Stoke's lines*. The lines with frequency greater than Rayleigh line ($\upsilon > \upsilon_0$) are called as *anti-Stoke's lines*. This effect is shown in Fig. 7.1. The Raman scattering is shown in Fig. 7.1.

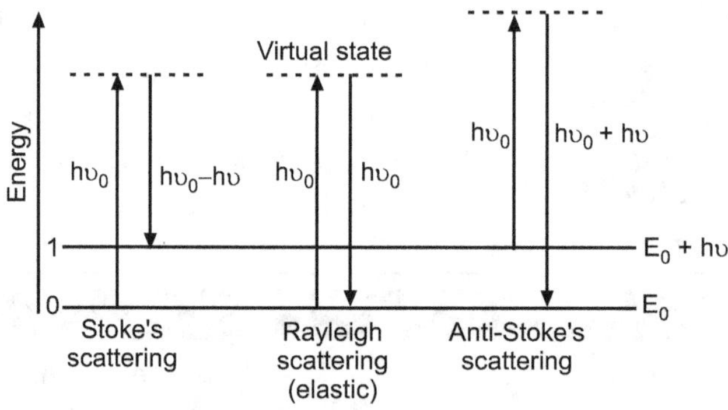

Fig. 7.1 : Raman effect

- In a transparent material, most of the molecules are initially in the ground state (state 0 in Fig. 7.1) but because of thermal agitation some molecules will be in an excited state (state 1). During the scattering process, the incoming photon raises the molecule to a virtual (i.e., non-existent) excited state. But the molecule cannot remain in this virtual level, so it immediately falls back down to a lower level with the emission of a photon. If the molecule falls into the same level from which it is excited, there is no frequency shift in the emitted photon and there is having Rayleigh scattering. If the molecule falls into a different level, the energy of the emitted photon must differ from that of the incoming photon in order to conserve total energy. In this case, the emitted photon has a different frequency than that of the incident photon. This process is called Raman scattering. The frequency of scattered photon may be less or more than incident photon. The decrease in frequency giving rise to lines in the spectrum called *Stokes lines* or increase in frequency give anti-Stokes lines depending on whether the molecule starts in the ground state or an excited state

7.1 Classical Theory of Raman Effect – Molecular Polarizability

- Phenomenon of Raman scattering can be understood on the basis of classical consideration. When a molecule is placed in an electric field, negative electron cloud is attracted towards positive pole, whereas positively charged nuclei get attracted towards negative pole and polarization of molecule takes place. Separation of charges induces dipole moment μ. The induced dipole moment μ is proportional to the applied electric field of intensity E.

$$\mu = \alpha E$$

- The constant of proportionality α is the molecular polarizability.
- When radiation of frequency υ_0 is allowed to fall on the molecule, each molecule experiences varying electric field.

$$E = E_0 \sin (2\pi\upsilon_0 t) \qquad \text{... (7.1)}$$

The induced dipole moment undergoes oscillation of frequency υ_0.

$$\mu = \alpha E = \alpha E_0 \sin (2\pi\upsilon_0 t) \qquad \text{... (7.2)}$$

- Such oscillating dipole emits radiation of its own oscillating frequency. Thus equation (7.2) gives explanation of Rayleigh scattering.
- While deriving the Raman effect, it is generally easiest to start with the classical interpretation by considering a simple diatomic molecule as a mass on a spring (as shown in Fig. 7.2) where m represents the atomic mass, x represents the displacement, and K represents the bond strength.

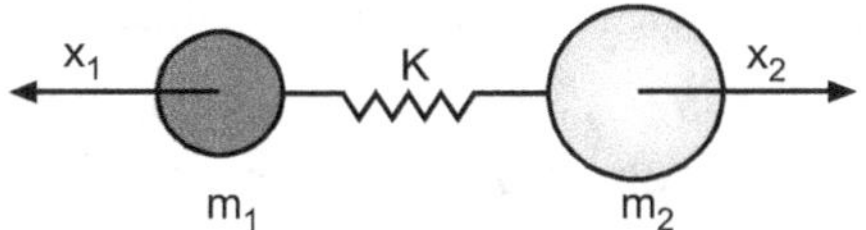

Fig. 7.2 : Vibrating diatomic molecule

- Let υ_{vib} be the frequency of vibration of the molecule, then the instantaneous displacement of the reduced mass m is given by

$$x = x_0 \sin (2\pi\upsilon_{vib} t)$$

where υ_{vib} is the frequency of molecular vibration and is defined as,

$$\upsilon_{vib} = \frac{1}{2\pi}\sqrt{\frac{K}{m}}$$

Thus due to internal motion of vibration or rotation, the polarizability changes periodically. Then oscillating dipole will have superimposed upon its vibrational or rotational oscillation.

Thus, we can write

$$\alpha = \alpha_0 + \beta \sin (2\pi \upsilon_{vib} t) \qquad \qquad ...(7.3)$$

where α_0 is equilibrium polarizability and β represents the rate of change of polarizability with vibration.

Then equation (7.3) becomes

$$\mu = \alpha E = (\alpha_0 + \beta \sin 2\pi \upsilon_{vib} t) \, E_0 \sin 2\pi \upsilon_0 t \qquad \qquad ...(7.4)$$

Using trigonometric relation,

$$\sin A \sin B = \frac{1}{2} \{\cos (A - B) - \cos (A + B)\}$$

Equation (7.4) becomes

$$\mu = \alpha_0 \, E_0 \sin (2\pi \upsilon_0 \, t) + \frac{1}{2} \beta E_0 \{\cos (2\pi \, (\upsilon_0 - \upsilon_{vib}) \, t) - \cos (2\pi \, (\upsilon_0 + \upsilon_{vib}) \, t)\}$$

Thus oscillating dipole has frequency component $\upsilon \pm \upsilon_{vib}$ as well as executing frequency υ.

(i)　　$\upsilon = \upsilon_0$　　　　　　　Rayleigh line

(ii)　　$\upsilon = \upsilon_0 - \upsilon_{vib}$　　　　Raman Stoke's line

(iii)　$\upsilon = \upsilon_0 + \upsilon_{vib}$　　　　Raman anti-Stoke's line

7.2 Quantum Theory of Raman Effect　　　(April 17, 16; Oct. 17, 15)

- Quantum mechanical picture is based on Kramer-Heisenberg theory of light. According to this theory, when a photon of frequency υ_0 interacts with molecule of substance, it will jump from any of its initial vibration or rotational level $E'_{v'/j'}$ (where v'/j' = 0, 1, 2, 3, ...) to any higher vibrational or rotational level $E''_{v''/j''}$. The higher energy level is unstable level for photon and molecule also. Its life time is of the order of 10^{-17} sec. The molecule in the virtual level subsequently returns to any one of the initial vibrational and rotational level $E'_{v'/j'}$ with emission of photon. The process of release of photon may be divided into the following three cases :

Case I :

The molecule may return from virtual state to any of its initial state such that the frequency of released photon is that of incident photon.

i.e.　　　　　　$E''_{v''/j''} - E'_{v'/j'} = h\upsilon_0 \qquad \qquad ...(7.5)$

or　　　　　　　　$\upsilon_0 = \dfrac{E''_{v''/j''} - E'_{v'/j'}}{h}$

This is true for $0 \to 0$, $1 \to 1$, $2 \to 2$, etc. transitions. This phenomenon of emission of photon is called as Rayleigh scattering and is due to the interaction of photon with nucleus. The energy level diagram of Rayleigh scattering is shown in Fig. 7.3 (a). This indicates the presence of unmodified Rayleigh line in the scattered light Raman scattering.

Case II :

The molecule in state $v'/j' = 0$ absorbs incident photon and jumps to virtual level and then subsequently returns to any one of the initial states with the release of photon with lower frequency (v^s) than the incident photon i.e.

$$E''_{v''/j''} - E'_{v'/j'} = hv^s \qquad \qquad \ldots (7.6)$$

This is true for $0 \to 1$, $0 \to 2$, $0 \to 3$, etc. transitions.

This is known as Stoke's Raman scattering. The diagram for Stoke's Raman scattering or effect is as shown in Fig. 7.3 (b).

Case III :

The molecule in higher initial level $v'/j' = 1$ gains energy from incident photon and jumps to any one of the virtual levels. The molecule subsequently returns to any of the initial level with the release of the photon :

(i) With frequency v^a higher than that of incident photon i.e.

$$E''_{v''/j''} - E'_{v'/j'} = hv^a \qquad \qquad \ldots (7.7)$$

This is true for $1 \to 0$, $2 \to 0$, etc. transitions. This process of release of photon is called anti-Stoke's Raman effect.

(ii) With lower frequency (v^s) by mechanism of case II. This will be true for $1 \to 2$, $1 \to 3$, etc. transitions and shown in Fig. 7.3 (c).

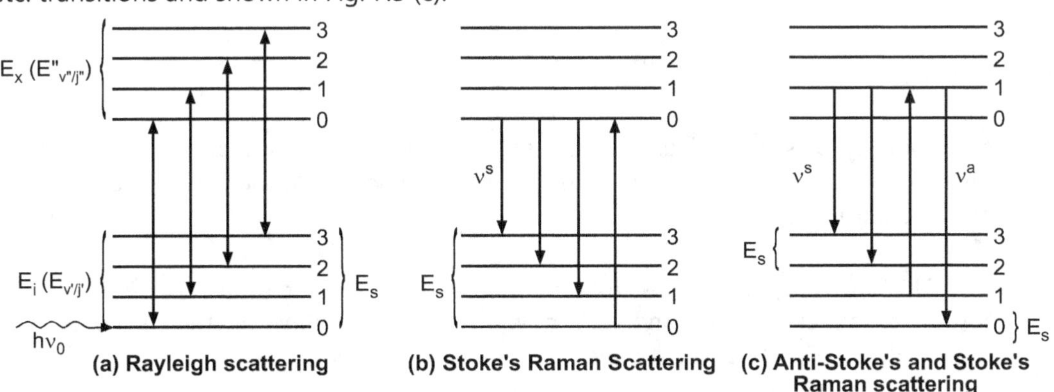

Fig. 7.3 : Energy level diagrams of Raman effect

- According to quantum theory, a wave of frequency v_0 consists of photon of each of energy hv_0. When this light falls on molecules of solid, it undergoes a collision with the molecule.

1. If the collision is elastic, there is no transfer of energy from the photon to molecule and vice versa. Hence photon is scattered without change in energy and frequency of the photon is same as incident photon. This explains the presence of Rayleigh line.

2. If the collision is elastic, there is exchange of energy between photon and molecule.

Let the incident photon with energy $h\upsilon$ collide with a molecule of kinetic energy $\frac{1}{2}mv^2$ and intrinsic energy E'. Let υ^s be the frequency of scattered photon, v_1 be velocity of molecule and E" intrinsic energy, then according to conservation of energy,

$$h\upsilon_0 + \frac{1}{2}mv^2 + E' = h\upsilon_1 + \frac{1}{2}mv_1^2 + E"$$

As there is no change in temperature, there is no change in K.E.

$$\frac{1}{2}mv^2 = \frac{1}{2}mv_1^2$$

Hence,
$$h\upsilon_0 + E' = h\upsilon^s + E"$$
$$h\upsilon_0 = h\upsilon^s + (E" - E')$$
$$h\upsilon_0 = h\upsilon^s + \Delta E$$
$$\upsilon_0 = \upsilon^s + \frac{\Delta E}{h}$$

There are two cases :

Case I :

Molecule gains energy from the photon and goes from a lower energy E' to a higher energy state E".

$\therefore$
$$E" - E' = \Delta E \text{ is positive}$$
$$\upsilon_0 = \upsilon^s + \frac{\Delta E}{h}$$

Or
$$\upsilon^s = \upsilon - \frac{\Delta E}{h}$$
$$\upsilon^s = \upsilon - \Delta\upsilon$$

- The frequency of scattered photon is less than the incident photon. This explains the presence of modified Stoke's line on lower frequency or longer wavelength side of Raman spectrum.

Case II :

The molecule may be at excited state initially. It may lose energy to the photon, which may now have frequency υ^a. The molecule now goes to initial higher energy level E' to lower energy level E", so that E" – E' is negative.

Hence
$$\upsilon_0 = \upsilon^a - \frac{\Delta E}{h}$$
$$\upsilon^a = \upsilon_0 + \frac{\Delta E}{h}$$

or
$$\upsilon^a = \upsilon_0 + \Delta\upsilon$$

The frequency of scattered photon is higher than that of incident photon. This explains presence of anti-Stoke's lines on the higher frequency or shorter wavelength side of the Raman spectrum.

Characteristics of Raman Lines :

The important parameters of Raman lines are :

1. **Intensity :**

   Intensity of Raman line varies directly as $(v_o - v_{vib.})^4$ except in vicinity of Raman Resonance Scattering i.e. intensity depends on frequency of radiation.

   Stoke's lines are invariably stronger than anti-Stoke's lines i.e. anti-Stoke's lines are generally fainter than Stoke's lines.

   Intensity of anti-Stoke's lines grows with temperature.

2. **Width and shape :**

   Raman lines are narrow in gases, liquids and crystals; while in amorphous substance, lines are broad and diffuse. Width of Raman line increases with increase in temperature. The shape of Raman line is independent of incident radiation.

3. **Frequency shift :**

   Frequency shift of Raman line from Rayleigh line varies from 4000 to few wave number and independent of frequency of incident radiation.

4. **Polarisation :**

   Raman scattered radiation is at least partially polarised irrespective of the polarisation of the incident radiation. Polarisation depends on structure of molecule as well as symmetry of vibrational motion. The scattered radiation may be polarised in a plane parallel and perpendicular to the plane of polarisation.

   Let $I_{\parallel}$ and $I_{\perp}$ be intensity of scattered radiation polarised in parallel and perpendicular planes respectively.

   The ratio of $I_{\parallel}$ to $I_{\perp}$ is called as degree of depolarisation and denoted by

$$\rho = \frac{I_{\parallel}}{I_{\perp}}$$

Depolarisation plays important role in determining the symmetry of a vibrational mode.

Theoretically, ρ is less than 3/4 for totally symmetric modes.

If $\rho = 3/4$, modes are not totally symmetrical.

If $\rho < 6/7$, then vibrations are asymmetric and capable of polarisation of radiation.

When $\rho > 6/7$, depolarisation of radiation occurs.

7.3 Experimental Setup to Observe Raman Spectra   (Oct. 17, 15)

- The Raman effect has been studied by a large number of workers. The general technique used by these researchers is to illuminate the substance under investigation with high monochromatic source of light, and photograph of scattered radiation is taken by arranging spectrograph in transverse direction.

- The original simple arrangement of Raman was not quite efficient and required very long exposure of about 100 hours and more to obtain good records of the Raman spectrum. Hence, improvements were made as regards the container of the substance, the source of radiation, filter and spectrograph, etc.

- The first apparatus used was developed by Wood and ordinarily used to study Raman effect in liquid. Certain modifications in the experimental arrangement were done for study of Raman effect in solid and liquid.

- To obtain a good Raman spectra, the incident beam should be strong as Raman scattering is weak. Secondly, the beam should be highly monochromatic as Raman lines are observed on both sides of central line (Rayleigh line). This purpose is served by using Laser. The complete flow sheet diagram of basic Raman spectrometer is shown in Fig. 7.4.

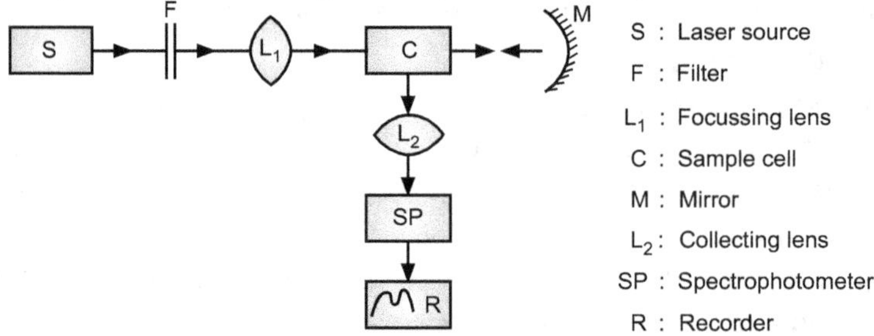

Fig. 7.4 : Schematic diagram of a simple Raman spectrometer

Raman Spectrometer consists of

(i)   a suitable gas laser source S and

(ii)   an optical system comprising filter F, focussing lens L_1 and mirror M to focus the beam on or into sample. There is a collecting lens L_2 to feed the scattered radiation to spectrophotometer SP. Spectrophotometer counts the photons as a function of frequencies. This analysis is displayed as a plot of scattering intensity versus frequency by the recorder R. Typical hypothetical spectra is represented in Fig. 7.5.

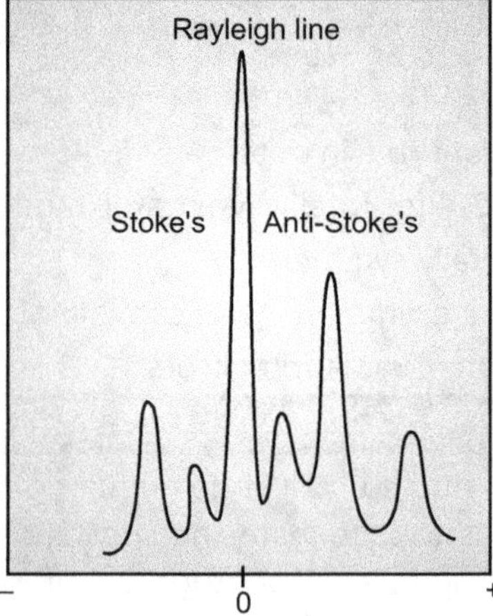

Fig. 7.5 : Hypothetical Raman spectrum using laser

1. Laser Source :

Before the discovery of laser, mercury arcs were used as a source of radiation.

The advent of accessible and relatively inexpensive laser sources during the past few years has caused major revolution in Raman scattering by displacing the traditional mercury source (discharge tube) as an exciting source. Now-a-days, a laser is ideal Raman source. It gives narrow, highly monochromatic beam of radiation which can be focussed into sample through filter.

The commonly used laser sources are :

1. Continuously operating He - Ne (633 nm), Ar^+ ion (488 nm) and Hg - Cd (441, 325 nm) gas lasers.

2. Pulsed and continuously operating forms of Ruby (694 nm) and Nd-YAG (1.06 micron) solid lasers.

3. Tunnable dye laser in visible range (0.43 – 0.65 μm).

4. UV laser (184 – 260 nm).

The power of laser is of the order of 50 mW – 2 W. The type of laser to be used depends upon the type of study i.e. He - Ne (100 mW, 633 nm) laser is frequently used due to narrow wavelength.

The advantages of using laser are as follows :

1. We do not need to cool the sample and need not use filter to get monochromatic beam of radiation.

2. Raman spectra can be recorded by using small amount of analyte.

3. Second order Raman spectra can be recorded.

4. The polarisation of laser beam is well defined and may be controlled within 0.1 %.

5. Broadening due to Doppler effect can be minimized.

6. As width of laser line is of the order of 0.005 cm^{-1} or less, a precise information could be obtained.

2. Sample Cell :

The sample cell is made up of non-fluorescent fused quartz having well polished flat bottom. The top of cell is sealed with a transparent cap to avoid dust and designed to permit deoxygenation of the sample. The cell is completely filled to avoid reflections. The spectra of liquid (coloured or colourless) is completely recorded by focussing the beam vertically through bottom window into the sample and then the scattered radiation perpendicular to incident beam is collected. The spectra of solid are recorded by using capillary. The thin capillary tube, filled with sample and sealed at one end, with the laser beam directed along its length. Radiation scattered from the sample is directed, via mirror, into a spectrometer operating in visible region. The spectrometer has monochromatic either a quartz or grating and detected by photoelectric detector. The output of detector is supplied to an amplifier and recorder (R). It is observed that solid gives more Rayleigh background. The spectra of sample, which are sensitive to laser wavelength, are recorded using rotating cell technique to avoid degradation of sample (like CCl_3 $COCl$ etc.).

In case of biological substance, local heating may degrade some biological molecule (i.e. it absorbs the energy at laser wavelength). In these type of biological systems, (e.g. cytochrome - C, haemoglobin) the resonance Raman scattering of laser radiation from vibrational mode of biological molecules are recorded. The hypothetical Raman spectrum is shown in Fig. 7.5.

7.4 Applications of Raman Spectroscopy   (April 17, Oct. 15)

Following are the important applications of Raman effect.

1.   Molecular structure : Raman spectra is useful for the study of structure of molecules. For this purpose we make an evaluation of wavelength of Raman lines, their intensities and state of polarization. Raman spectra include vibrational spectra of molecules. Natural frequency of a diatomic molecule is

$$\upsilon = \left(\upsilon + \frac{1}{2}\right)\upsilon_0$$

where
$$\upsilon_0 = \frac{1}{2\pi}\sqrt{\frac{k}{m'}}$$

$$k = \text{Restoring force per unit displacement}$$

and
$$m' = \text{Reduced mass of molecule}$$

From above relation for large k value, binding force between atoms is large. It has higher characteristic frequency than that one in which force is weak. The force depends upon nature of interatomic bonds. In a molecule having a covalent bond, the polarizability changes considerably due to nuclear oscillations, which give rise to intense Raman line. In a molecule having electrovalent bond, the binding electrons change over from one nucleus to other in the formation of molecule. The polarizability of a molecule is not effected by nuclear oscillations, hence no Raman line appears.

In case of triatomic molecule, from a study of Raman effect we can find whether the molecule is symmetric or asymmetric; linear or non-linear. From the number and intensity of observed lines in Raman effect along with data of infrared (IR) spectrum, important conclusion regarding molecular structure can be drawn.

Raman spectrum tells us that water (H_2O) molecule is though symmetric is non-linear or bent. The bend in system is about 120°.

2.   Nature of the liquid state : Observation of Raman spectra of liquid led to important conclusion regarding liquid state viz. that is characterized by both types of thermal energy, one in which it is organized in manner of elastic waves and other in which it is a kin to random motion.

3.   Crystal physics : Raman spectroscopy offers new method of studying crystals which is complementary to X-ray diffraction. It furnishes us the information that X-ray do not give concerning the strength and nature of forces which hold the crystal together. This information gives precise picture of crystal as Raman lines are very sharp and can be measured very accurately.

Raman effect in crystal has been studied with reference to other problems of crystal physics, viz. the existence of a mosaic structure and thermal agitation in crystals. These studies show the existence of discrete monochromatic IR frequencies which explain fine structure of IR spectrum of crystal.

4.   Nuclear physics : Raman spectroscopy has been applied in the study of some aspect of nuclear physics such as spin, statistics as well as isotopic constitution of nucleus. It is also useful in the study of rotational Raman lines of homonuclear diatomic molecules like H_2, D_2 (deuterium), O_2, N_2, etc. characterized by their alternating intensities. For example, for hydrogen, Raman lines for transitions between odd rotational levels are three times more intense than those arising from transition between even rotational levels.

5.   Chemistry : Interesting problems occurring in physical chemistry such as transition from crystalline to amorphous state, electrolytic dissociation, hydrolysis have been investigated with Raman effect. Raman effect has great value in organic chemistry to determine presence or absence of linkage in a molecule, identifying impurities of certain types. Various chemical effects like strength of chemical bond, electrolytic dissociation, hydrolysis have been understood through Raman effect.

6.   Biological applications :

-    In biology, Raman effect provides information on peptide backbone disulphide bridge and environment of some of side chains (tyrosine, hystidine, etc.)

-    It is useful in the study of structural and conformational changes and denaturation of macromolecule like protein and carbohydrates. It is also used to study structural and conformational aspect of lipids, biological membrane and lipid protein complex.

Solved Problems

Problem 7.1 : *A compound is irradiated by 4358 A° line of mercury. Raman lines are observed at wavelengths 4420 and 4620 A°. Compute the value of Raman shift for each line in terms of wave number.*   **(Oct. 16)**

Solution :

$$\upsilon_O - \upsilon_R = \Delta\upsilon_R$$

$$\frac{c}{\lambda_O} - \frac{c}{\lambda_R} = \frac{c}{\Delta\lambda_R}$$

or

$$\frac{1}{\lambda_O} - \frac{1}{\lambda_R} = \frac{1}{\Delta\lambda_R}$$

For λ = 4420 A°,

$$\frac{1}{4358 \times 10^{-8}} - \frac{1}{4420 \times 10^{-8}} = \overline{\Delta\upsilon}$$

$\therefore$ $\qquad\qquad\qquad\qquad \overline{\Delta\upsilon}$ = **321.9 cm^{-1}**   **... Ans.**

For λ = 4620 A°,

$$\overline{\Delta\upsilon} = \frac{1}{4358} - \frac{1}{4620} = \textbf{1301.3 cm}^{-1} \qquad \textbf{... Ans.}$$

Problem 7.2 : *A sample was excited by the 4358 A° line of mercury. A Raman line was observed at 4447 A°. Calculate Raman shift.*

Solution : Since Raman line is observed at longer wavelength (shorter frequency) than exciting line, it is a Stoke's line in Raman spectroscopy.

Raman shift : $\qquad\qquad \Delta\upsilon \ (cm^{-1}) = \dfrac{10^8}{\lambda_{exc} \ (in \ A°)} - \dfrac{10^8}{\lambda_{Raman} \ in \ (A°)}$

$$\Delta\upsilon = \frac{10^8}{4.358 \times 10^3} - \frac{10^8}{4.447 \times 10^3}$$

$$= (2.295 - 2.249) \times 10^4$$

$$= 0.046 \times 10^4$$

$$= \textbf{460 cm}^{-1} \qquad\qquad \textbf{... Ans.}$$

Problem 7.3 : *At what wavelength in A° would the anti-Stoke's line appear in the Raman spectrum of Problem 7.2.*

Solution : From previous example, Raman shift $\Delta\upsilon$ = 460 cm^{-1}

Anti-Stoke's line will appear at frequency 460 cm^{-1} higher than the frequency (in cm^{-1}) associated with 4358 A° Hg line (used as source of excitation).

$$\upsilon_{exc} \ (cm^{-1}) = \frac{10^8}{\lambda_{exc}} = \frac{10^8}{4.358 \times 10^3}$$

$$= 2.295 \times 10^4 \ cm^{-1}$$

$\therefore$ $\qquad\qquad \upsilon_{anti\text{-}Stoke} = (2.295 \times 10^4 \ cm^{-1}) + (460 \ cm^{-1})$

$$= 2.341 \times 10^4 \ cm^{-1}$$

$$\upsilon = \frac{c}{\lambda} = \frac{10^8}{\lambda}$$

$$\upsilon_{(in \ A°)} = \frac{10^8}{\upsilon \ cm^{-1}} = \frac{10^8}{2.341 \times 10^4} = \textbf{4272 A}° \qquad \textbf{... Ans.}$$

Problem 7.4 : *In an experiment of Raman effect using mercury green radiation of λ = 546.1 nm, a Stoke's line of wavelength 554.3 nm was observed. Find Raman shift and wavelength corresponding to anti-Stoke's line.*

Solution :　　　　$\lambda = 546.1$ nm $= 5461 \times 10^{-10}$ m

Frequency of exciting radiation

$$\upsilon_0 = \frac{c}{\lambda} = \frac{c}{5461 \times 10^{-10}} = c \times 1.8312 \times 10^6 \text{ s}^{-1}$$

Wavelength of Stoke's line,

$$\lambda_s = 554.3 \text{ nm} = 5543 \times 10^{-10} \text{ m}$$

∴　Frequency of Stoke's line

$$\upsilon_s = \frac{c}{\lambda_s} = \frac{c}{5543 \times 10^{-10}} = c \times 1.8041 \times 10^6 \text{ s}^{-1}$$

∴　　　　Raman shift $= \upsilon - \upsilon_s$

$$= c \,(1.8312 \times 10^6 - 1.8041 \times 10^6) = 3 \times 10^8 \times 2.71 \times 10^4$$

$$= \mathbf{813 \times 10^{-10} \text{ s}^{-1}} \qquad\qquad \textbf{... Ans.}$$

∴　Frequency of anti-Stoke's line (i.e. $\upsilon_a > \upsilon_o$)

$$\upsilon_a = \upsilon_o + c \times 2.71 \times 10^4$$

$$= \mathbf{c \times 10^2 \,(18312 + 271)} \qquad\qquad \textbf{... Ans.}$$

Wavelength of anti-Stoke's line,

$$\lambda_a = \frac{c}{c \times 10^2 \times 18583} = 5.381 \times 10^{-5} \times 10^{-2}$$

$$= \mathbf{538.1 \text{ nm}} \qquad\qquad \textbf{... Ans.}$$

Summary

1. When electromagnetic radiation passes through pure transparent substance, it undergoes scattering giving central line of frequency υ_o (called Rayleigh line) and series of displaced lines with frequency υ_R). This phenomenon of scattering with modified frequency is called Raman scattering.

2. Raman scattering is weak and about 1% of the incident radiation.

3. When a molecule is placed in an electric field, negative electron cloud is attracted towards positive poles, positively charged nuclei get attracted towards negative pole and polarization of medium takes place. The dipole moment μ so induced is proportional to the applied electric field.

$$\mu = \alpha E$$

Constant of proportionality α is called molecular polarizability.

4. Raman effect is well explained by quantum theory.

5. Intensity of Raman line varies directly as $(\upsilon_o - \upsilon_{vib})^4$.

6.　Raman lines are narrow in gases, liquids and crystals, but in amorphous substance lines are broad and diffused.

7.　Raman spectroscopy is useful in

(a)　understanding molecular structure

(b)　study of liquid state

(c)　crystal physics

(d)　nuclear physics

(e)　chemistry and biology.

Exercise

(A) Short Answer Type Questions :

1.　What is Raman effect ?

2.　What is molecular polarizability ?

3.　What are Stoke's lines ? What are anti-Stoke's lines ?

4.　What is Rayleigh line ?

5.　What is Raman shift ?

6.　What are the applications of Raman effect in physics ?

(B) Long Answer Type Questions :

1.　What is Raman effect ? With proper sketch explain occurrence of Stoke's lines and anti-Stoke's lines.

2.　Explain classical theory of Raman effect.

3.　Explain Raman effect on the basis of quantum theory.

4.　Describe the experimental arrangement to observe Raman spectra.

5.　What are the applications of Raman spectra ?

(C) Numerical Problems :

1.　The Rayleigh line in Raman experiment is 4358 A° and the spectrum shows the Stoke's line at 6433 A°. Find the wavelength of anti-Stoke's line.　　**(Ans.** 3295 A°)

2.　A substance shows Raman line at 4567 A° when exciting line 4358 A° is used. At what position Stoke's and anti-Stoke's lines for the same substance will be observed when the position of Rayleigh line is at 4047 A° ?

(Ans. 4227 A° (Stoke's line), 3882 A° (anti-Stoke's line))

3.　The exciting line in an experiment with Raman effect is 5460 A°. If Stoke's line has λ = 5520 A°, calculate the wavelength of anti-Stoke's line.　　**(Ans.** λ = 5400 A°)

1. **Attempt all of the following (one mark each) :**   **(10)**

   (a)   What is the meaning of space quantization ?

   Ans. Refer to Section 1.5 on page 1.15.

   (b)   What is multiplicity ? Calculate the value of S for triplet.

   Ans. Refer to Section 2.1.4 on page 2.7.

   (c)   State Hund's rule for electronic configuration of an atom.

   Ans. Refer to Section 2.1.2 on page 2.3.

   (d)   Determine all possible values of L for two D electrons ($l_1 = 2$, $l_2 = 2$) using jj coupling.

   Ans. Refer to Section 3.1.2 on page 3.5.

   (e)   What is Stark effect ?

   Ans. Refer to Section 4.5 on page 4.13.

   (f)   State Moseley's law.

   Ans. Refer to Section 5.4 on page 5.9.

   (g)   What do you mean by fine structure of X-ray lines ?

   Ans. Refer to Section 5.5 on page 5.11.

   (h)   Define fluorescence.

   Ans. Refer to Section 6.4 on page 6.14.

   (i)   Give the selection rule for transition between vibrational energy levels.

   Ans. Refer to Section 6.2 on page 6.11.

   (j)   Give applications of Raman effect in Nuclear physics.

   Ans. Refer to Section 7.4 on page 7.10.

2. **Attempt any Two of the following :**

   (a)   What are X-rays ? Explain the origin of characteristic X-ray spectra.   **(5)**

   Ans. Refer to Section 5.1 on page 5.3. Also refer to Section 5.3 on page 5.7.

   (b)   Obtain an expression for rotational energy level of rigid diatomic molecule.   **(5)**

   Ans. Refer to Section 6.1 on page 6.3.

   (c)   What is Raman effect ? Discuss the experimental setup of Raman spectrometer with neat diagram.   **(5)**

   Ans. Refer to Introduction on page 7.1. Also refer to Section 7.3 on page 7.7.

3. **Attempt any Two of the following :**

   (a)   The separation of Zeeman components of 600 nm spectral line are 0.015 nm, when the magnetic field is 1.20 Tesla. Calculate the ratio e/m for the electron.

       [Given : $c = 3 \times 10^8$ m/s]   **(5)**

   Ans. Refer to Problem 4.3 on page 4.16.

(b)   The bond length of carbon monoxide (CO) molecule is 0.113 nm and the masses of 12C and 16O atoms are respectively 1.99×10^{-26} kg and 2.66×10^{-26} kg. Calculate the reduced mass, moment of inertia and rotational energy in electron volt in its rotational state J = 1. **(5)**

[Given : 1 eV = 1.6×10^{-19} J, h = 6.63×10^{-34} J.s]

Ans. Refer to Problem 6.1 on page 6.15.

(c)   Derive all the atomic terms arising from p.d. configuration using jj coupling scheme and draw the vector diagrams. **(5)**

Ans. Refer to Section 3.1.2 on page 3.5.

4. (a)   Attempt any One of the following :

(i)   What are quantum numbers ? Explain all four quantum numbers necessary to define the quantum state of an electron. **(8)**

Ans. Refer to Section 1.5 on page 1.15. Also refer to Section 1.5 on page 1.19.

(ii)   What is spin-orbit interaction energy ? Obtain an expression. **(8)**

$$\Delta W_{ls} = \frac{Ze^2}{2m^2c^3}\frac{h^2}{4\pi^2} \cdot \frac{1}{r^3}\, l^*\, s^* \cos(l^*, s^*)$$

Where symbols have their usual meaning.

Ans. Refer to Section 2.2.1 on page 2.8.

(b)   Attempt any One of the following :

(i)   State Lande's interval rule. **(2)**

Ans. Refer to Section 3.3 on page 3.13.

(ii)   Explain Normal Zeeman effect. **(2)**

Ans. Refer to Section 4.3 on page 4.6.

❏❏❏

April 2016

1. Attempt all of the following (one mark each) : **(10)**

(a)   Determine the values of J for S = 1 and L = 3.

Ans. Refer to Section 2.1.4 on page 2.7.

(b)   What is electronic configuration ? Give the electronic configuration of oxygen atom (Z = 8).

Ans. Refer to Section 2.1.2 on page 2.3.

(c)   Define quantum state of an electron.

Ans. Refer to Section 2.1.3 on page 2.6.

(d)   What are X-rays ?

Ans. Refer to Section 5.1 on page 5.3.

(e)   What do you mean by pure rotational spectrum of a molecule ?

Ans. Refer to Section 6.1 on page 6.5.

(f)   What are stoke's and anti-stoke's lines in Raman spectra ?

Ans. Refer to Section 7.2 on page 7.4.

(g)   State Bohr's second postulate in Bohr's theory.

Ans. Refer to Section 1.3 on page 1.8.

(h)   Give any two limitations of Bohr's theory.

Ans. Refer to Section 1.4 on page 1.11.

(i)   What is the selection rule for L for allowed transition ?

Ans. Refer to Section 2.2.3 on page 2.15.

(j)   Define fluorescence.

Ans. Refer to Section 6.4 on page 6.14.

2. **Attempt any Two of the following :**

(a)   What is Raman effect ? Discuss the Raman effect on the basis of quantum theory.   **(5)**

Ans. Refer to Introduction on page 7.1. Also refer to Section 7.2 on page 7.4.

(b)   Explain the concept of space quantization and draw the vector diagram of space quantization of orbital angular momentum for $l = 4$.   **(5)**

Ans. Refer to Section 1.5 on page 1.15.

(c)   Explain normal Zeeman effect for single valence electron system and show that

$$\upsilon = \upsilon_0 + \frac{eH}{4\pi m}\Delta m_l$$ where, symbols have their usual meaning.   **(5)**

Ans. Refer to Section 4.4 on page 4.10.

3. **Attempt any Two of the following :**

(a)   An electron collides with a hydrogen atom in its ground state and excite it to a state of n = 3. Calculate the energy given by the electron to hydrogen atom during the inelastic collision. (Given : mass of electron m = 9.1×10^{-31} kg, e = 1.6×10^{-19} C, h = 6.63×10^{-34} J.s, $\in_0 = 8.85 \times 10^{-12}$ C^2/N-m^2)   **(5)**

Ans. Refer to Problem 1.2 on page 1.27.

(b)   The mean of the intermolecular distances for HCl molecule in the v = 0 and v = 1 level is 1.30 A°. The masses of hydrogen atom and chlorine atom are 1.7×10^{-24} gm and $35 \times 1.7 \times 10^{-24}$ gm respectively. Calculate the moment of inertia and rotational constant of HCl molecule.   **(5)**

(Given : h = 6.63×10^{-27} erg. sec, c = 3×10^{10} cm/s).

Ans. Refer to Problem 6.3 on page 6.16.

(c)   If an electron in a tungsten atom (Z = 74) jumps from M level to L level and given Lα line in L series of X-rays, calculate the wavelength of emitted Lα X-ray line, if nuclear screening constant is 6.20.   **(5)**

(Given : R = 1.097×10^7 m^{-1})

Ans. Refer to Problem 5.4 on page 5.17.

4. (A)   Attempt One of the following :   **(8)**

(a)   Determine the ground state and excited atomic states of sodium atom. Represent various states using spectral notations and discuss the spectra of sodium atom with the help of energy level diagram and selection rules.

Ans. Refer to Section 2.2.4 on page 2.17.

(b)   Determine singlet-triplet separations in terms of interaction energies between two valence electrons in 'pd' configuration using L-S coupling. Draw the fine structure energy level diagram.   **(8)**

Ans. Refer to Section 3.2 on page 3.7.

(B)   Attempt One of the following :

(a)   What are three major types of molecular spectra ?   **(2)**

Ans. Refer to Introduction on page 6.2.

(b)   State Pauli's exclusion principle and Hund's rule.   **(2)**

Ans. Refer to Sections 2.1.1 and 2.1.2 on page 2.2 and 2.3.

❏❏❏

October 2016

1.   Attempt all of the following :   **(10)**

(a)   Define the term : Multiplicity of state.

Ans. Refer to Section 2.1.4 on page 2.7.

(b)   Give the electronic configuration for sodium atom (Z = 11).

Ans. The electronic configuration for sodium atom is $1s^2, 2s^2, 2p^6, 3s^1$.

(c)   State Pauli's exclusion principle.

Ans. Refer to Section 2.1.1 on page 2.2.

(d)   Define the reduced mass of diatomic molecule.

Ans. Refer to Section 1.3 on page 1.6.

(e)   State second postulate of Bohr's theory for hydrogen atom.

Ans. Refer to Section 1.3 on page 1.8.

(f)   Determine the values of Me for l = 5.

Ans. Refer to Section 2.1.3 on page 2.6.

(g)   State Duane and Hunt's law in X-rays.

Ans. Refer to Section 5.2 on page 5.4.

(h)   Write the atomic states for L = 2 and S = $\frac{1}{2}$.

Ans. Refer to Problem 2.3 on page 2.20.

(i)   Define the term : Larmour precession.

Ans. Refer to Section 3.2 on page 3.7.

(j)   Give the quantum state of an electron.

Ans. Refer to Section 2.1.3 on page 2.6.

2. **Attempt any two of the following :**   (10)

(a)   What is normal Zeeman effect ? Derive an expression $v = v_0 + \Delta m_l \dfrac{eH}{4\pi m}$ where symbols have their usual meanings.

Ans. Refer to Section 4.3 on page 4.7.

(b)   Explain the origin of continuous X-rays.

Ans. Refer to Section 5.2.1 on page 5.5.

(c)   Obtain an expression for vibrational energy levels of diatomic molecule.

$$E_v = \left(v + \frac{1}{2}\right)\frac{h}{2\pi}\sqrt{\frac{k}{m}}$$ where symbols have their usual meanings.

Ans. Refer to Section 6.2 on page 6.7.

3. **Attempt any two of the following :**   (10)

(a)   Determine the ground state of aluminium atom (Z = 13) and represent it using spectral notations.

Ans. Refer to Problem 2.2 on page 2.20.

(b)   Find the linear velocity of an electron in first and fourth orbit of hydrogen atom.
      Given : $E_0 = 8.85 \times 10^{-12}$ C^2/Nm^2, h = 6.64×10^{-34} Js,
                       e = 1.6×10^{-19} C

Ans. Refer to Problem 1.3 on page 1.27.

(c)   A sample of compound is irradiated with mercury source of wavelength 4358 A°. Raman lines are observed at wavelengths 4620 A° and 4420 A°. Determine the value of Raman shift in wavenumber for each line.

Ans. Refer to Problem 7.1 on page 7.11.

4. (a) **Attempt any one of the following :**   (8)

(i)   What is L-S coupling ? Obtain the spectral terms for two valence electron atom in p-d configuration. Draw the vector diagram.

Ans. Refer to Section 3.1.1 on page 3.2.

(ii)   What are quantum numbers ? Give their physical significance.

Ans. Refer to Section 2.1.1 on page 2.2. Also refer to Section 2.1.3 on page 2.6.

(b)   Attempt any one of the following :   (2)

(i)   What are stoke and antistoke lines ?

Ans. Refer to Section Introduction on page 7.1.

(ii)   State Moseley's law and one application of it.

Ans. Refer to Section 5.4 on page 5.9.

❏❏❏

April 2017

1. **Attempt all of the following :**   (10)

(a)   State Duane and Hunt's law in X-ray.

Ans. Refer to Section 5.2 on page 5.4.

(b)   What is normal Zeeman effect ?

Ans. Refer to Section 4.3 on page 4.6.

(c)   Give the electronic configuration for aluminium atom, Z = 13.

Ans. The electronic configuration for aluminium atom is $1s^2, 2s^2, 2p^6, 3s^2, 3p^1$.

(d)   Calculate the multiplicity for total spin S = 1.

Ans. Refer to Section 2.1.4 on page 2.7.

(e)   State Moseley's law.

Ans. Refer to Section 5.4 on page 5.9.

(f)   State the third postulate of Bohr's theory of hydrogen atom.

Ans. Refer to Section 1.3 on page 1.8.

(g)   What is Stark effect ?

Ans. Refer to Section 4.5 on page 4.13.

(h)   State three types of molecular spectra.

Ans. Refer to Section Introduction on page 6.2.

(i)   Give any two applications of Raman spectroscopy.

Ans. Refer to Section 7.4 on page 7.10.

(j)   Calculate the m values for $l = 3$.

Ans. Refer to Section 2.1.3 on page 2.6.

2. **Attempt any two of the following :**   (10)

(a)   State Pauli's exclusion principle and show that total number of states for given n are equal to $2n^2$.

Ans. Refer to Section 2.1.1 on page 2.2.

(b)   State the Lande's interval rule and represent it graphically for $3D_{12}$ terms.

Ans. Refer to Section 3.3 on page 3.13.

(c)   Explain the characteristic X-ray emission spectrum.

Ans. Refer to Section 5.3 on page 5.7.

3. Attempt any two of the following :   **(10)**

(a)   The separation of Zeeman components of 500 nm spectral line is 0.0106 nm when the magnetic field is 0.40 Tesla. Calculate the ratio e/m. Given : $c = 3 \times 10^8$ m/s.

Ans. Refer to Problem 4.3 on page 4.16.

(b)   The force constant of the bond in CO molecule is 1956 N/m. The reduced mass of CO molecule is 1.16×10^{-26} kg. Determine the frequency of vibration and spacing between its vibrational levels in eV.

Given : $h = 6.63 \times 10^{-27}$ erg. sec., 1 eV = 1.6×10^{-12} erg.

Ans. Refer to Problem 6.1 on page 6.15.

(c)   Determine the ground state of sodium atom (Z = 11) and represent it using the spectral notations.

Ans. Refer to Section 2.2.4 on page 2.17.

4. (a)   Attempt any one of the following :   **(8)**

(i)   Explain the space quantization and electron spin concepts. Draw the space quantization diagram for $l = 3$.

Ans. Refer to Section 2.1.4 on page 2.7.

(ii)   Obtain an expression for rotational energy levels of rigid diatomic molecule. Explain rotational spectra.

Ans. Refer to Section 6.1 on page 6.3.

(b)   Attempt any one of the following :   **(2)**

(i)   Explain Raman and Rayleigh lines.

Ans. Refer to Section 7.2 on page 7.4.

(ii)   State the four quantum numbers for quantum state.

Ans. Refer to Section 2.1.4 on page 2.7.

❑❑❑

October 2017

1. Attempt all of the following (one mark each) :   **(10)**

(a)   What are stoke's and antistoke's lines in Raman effect?

Ans. Refer to Section 7.2 on page 7.4.

(b)   Define reduced mass of a diatomic molecule.

Ans. Refer to Section 1.3 on page 1.6.

(c)　Give three types of molecular spectra.

Ans. Refer to Section Introduction on page 6.2.

(d)　State Duane and Hunt's law in X-rays.

Ans. Refer to Section 5.2 on page 5.4.

(e)　What is anomalous Zeeman effect ?

Ans. Refer to Section 4.3 on page 4.6.

(f)　Define multiplicity of an atomic state.

Ans. Refer to Section 2.1.4 on page 2.7.

(g)　Determine values of L of two electron system with $l_1 = 3$ and $l_2 = 2$.

Ans. Refer to Section 3.2 on page 3.7.

(h)　What is quantum state of an electron ?

Ans. Refer to Section 2.1.3 on page 2.6.

(i)　State Bohr's third postulate for hydrogen atom.

Ans. Refer to Section 1.3 on page 1.8.

(j)　Determine the values of S, L and J for the atomic state $^2D_{3/2}$.

2. **Attempt any Two of the following :**

(a)　What is normal Zeeman effect ? Obtain an expression for normal Zeeman effect in the form : $v = v_0 + \Delta m_l \dfrac{eH}{4\pi m}$ where symbols have usual meanings.　**(5)**

Ans. Refer to Section 4.3 on page 4.6.

(b)　Give the comparison of X-ray spectra and optical spectra.　**(5)**

Ans. Refer to Section 5.6 on page 5.13.

(c)　Discuss an experimental setup to observe the Raman effect.　**(5)**

Ans. Refer to Section 7.3 on page 7.7.

3. **Attempt any Two of the following :**

(a)　Calculate the linear velocities of an electron in first, fourth and tenth orbits of hydrogen atom.　**(5)**

$\varepsilon_0 = 8.85 \times 10^{-12}\ C^2/Nm^2$, $h = 6.64 \times 10^{-34}\ Js$, $e = 1.6 \times 10^{-19}\ C$.

Ans. Refer to Problem 1.3 on page 1.27.

(b)　Determine all the atomic terms for p-d electron system using L-S coupling and draw the necessary vector diagrams.　**(5)**

Ans. Refer to Section 3.1.1 on page 3.2.

(c)　Determine the ground state of Aluminium atom (Z = 13) and represent it in spectral notation.　**(5)**

Ans. Refer to Problem 2.2 on page 2.20.

4. (a)　Attempt any One of the following :

(i)　Show that the vibrational energy levels of diatomic molecule are given by an expression $E_v = \left(v + \dfrac{1}{2}\right)\dfrac{h}{2\pi}\sqrt{\dfrac{K}{\mu}}$. Discuss the vibrational spectra.　**(8)**

Ans. Refer to Section 6.2 on page 6.7.

(ii)　Discuss the different series in case of sodium spectra with energy level diagram. **(8)**

Ans. Refer to Section 1.4 on page 1.9.

(b)　Attempt any One of the following :

(i)　Give any two applications of Moseley's law.　**(2)**

Ans. Refer to Section 5.4 on page 5.9.

(ii)　Give any two limitations of Bohr's theory of hydrogen atom.　**(2)**

Ans. Refer to Section 1.4 on page 1.9.

❏❏❏

April 2018

1.　Attempt all of the following (one mark each) :　**(10)**

(a)　State any two limitations of Bohr's theory.

Ans. Refer to Section 1.3 on page 1.6.

(b)　State Pauli's exclusion principle.

Ans. Refer to Section 2.1.1 on page 2.2.

(c)　What is ground state of an electron ?

(d)　State the value of m_l for $l = 4$.

Ans. Refer to Section 1.5 on page 1.15.

(e)　State formula for wavelength of Balmer series.

(f)　What is the conclusion of Frank-Hertz experiment ?

Ans. Refer to Section 1.6 on page 1.25.

(g)　What is the physical significance of principal quantum number ?

Ans. Refer to Section 2.1.4 on page 2.7.

(h)　Define equivalent electrons.

(i)　State Bohr's first postulate in Bohr's theory.

Ans. Refer to Section 1.3 on page 1.7.

(j)　State formula for reduced mass of a molecule.

2.　Attempt any two of the following :

(a)　What is Raman effect ? Describe experimental set-up to observe Raman spectra.**(5)**

Ans. Refer to Section Introduction on page 4.1.

(b)   State and explain Lande interval rule. Represent it graphically for ^3D term.   **(5)**

Ans. Refer to Section 3.3 on page 3.13.

(c)   Obtain an expression for rotational energy level of rigid diatomic molecule.   **(5)**

Ans. Refer to Section 6.1 on page 6.1.

3.   **Attempt any two of the following :**

(a)   In an experiment of Raman effect using mercury green radiation of λ = 546.1 nm, a Stoke's line of wavelength 554.3 nm was observed. Find Raman shift and wavelength corresponding to anti-Stoke's line.   **(5)**

Ans. Refer to Problem 7.4 on page 7.12.

(b)   A sample of certain element is placed in 1 Tesla magnetic field and suitably excited. How far apart are the Zeeman components of the 5000 A° spectral lines of this element ? Given : e = 1.6×10^{-19} C, m = 9.11×10^{-31} kg, c = 3×10^8 m/s.   **(5)**

Ans. Refer to Problem 4.2 on page 4.16.

(c)   The force constant of the bond in CO molecule is 1956 N/m. Calculate the frequency of vibration of the molecule and the spacing between its vibrational energy levels in eV.   **(5)**

Given : h = 6.63×10^{-27} erg-sec, 1 eV = 1.6×10^{-12} erg.

$$c = 3 \times 10^8 \text{ m/s}, \ \mu = 1.16 \times 10^{-26} \text{ kg}$$

Ans. Refer to Problem 6.1 on page 6.15.

4.   **(a)**   **Attempt any one of the following :**

(i)   What are X-rays ? Discuss in detail production of characteristic X-ray spectra with energy level diagram.   **(8)**

Ans. Refer to Section 5.1 on page 5.3. Also refer to Section 5.3 on page 5.7.

(ii)   Obtain an expression for spin-orbit interaction energies for two valence electron system (LS coupling).   **(8)**

Ans. Refer to Section 3.1.1 on page 3.2.

(b)   **Attempt any one of the following :**

(i)   What are 'L' and 'S' quantum numbers corresponding to $^3D_2$?   **(2)**

Ans. Refer to Section 1.5 on page 1.19.

(ii)   What is vibrational-rotational spectrum ?   **(2)**

Ans. Refer to Section 6.3 on page 6.13.

❑❑❑